ATLAS OF MACHINED SURFACES

ATLAS OF MACHINED SURFACES

K.J.Stout
School of Manufacturing and
Mechanical Engineering,
University of Birmingham, UK

E.J. Davis
Department of Mechanical Engineering
and Manufacturing Systems,
Coventry Polytechnic, UK

P.J. Sullivan
School of Manufacturing and
Mechanical Engineering,
University of Birmingham, UK

CHAPMAN AND HALL

LONDON • **NEW YORK** • **TOKYO** • **MELBOURNE** • **MADRAS**

UK   Chapman and Hall, 11 New Fetter Lane, London EC4P 4EE

USA   Van Nostrand Reinhold, 115 5th Avenue, New York NY10003

JAPAN   Chapman and Hall Japan, Thomson Publishing Japan, Hirakawacho
            Nemoto Building, 7F, 1 – 7 – 11 Hirakawa-cho, Chiyoda-ku, Tokyo 102

AUSTRALIA   Chapman and Hall Australia, Thomas Nelson Australia,
            480 La Trobe Street, PO Box 4725, Melbourne 3000

INDIA   Chapman and Hall India, R. Seshadri, 32 Second Main Road, CIT East,
            Madras 600 035

First edition 1990

© 1990 K.J. Stout, E.J. Davis and P.J. Sullivan

Typeset in 11/13 Times by Wagstaffs Design, Bedford
Printed in Great Britain at the University Press, Cambridge

ISBN   0 412 37710 1   0442 31196 6 (USA)

British Library Cataloguing in Publication Data

Stout, K.J.

  1. Engineering components. Surfaces. Topography
  I. Title II. Davies, E.J. III. Sullivan, P.
  620

  ISBN 0–412–37710–1

Library of Congress Cataloging-in-Publication Data
Available

A software package, which is closely similar to the 3-D system offered on Rank Taylor
Hobson Form Talysurf series machines, is available containing all the data presented in Part
1 of this atlas. The package allows the user to manipulate the data to produce a full range of
projections of the type presented here, together with truncation, inversion and the associated
parameters. Additional features available include: zoom, grey-scale imaging and full colour
graphics.

  The package will run on any IBM AT compatible machine with VGA graphics.
  For details of the software package, please apply to:
    Whitestone Business Communications
    24 Golf Drive
    Whitestone
    Nuneaton
    Warwickshire
    CV11 6LY

CONTENTS

INTRODUCTION

A variety of manufacturing processes are used to create engineering surfaces, each of which produces a surface with its own characteristic topography. It is important to realize that this topography may affect the suitability of a surface for specific functional applications. Unfortunately, the relationship between surface topography and functional behaviour is not yet fully understood. It is clear, however, that there are two quite distinct issues which need to be addressed: (1) the relationship between manufacture and the resulting surface topography, and (2) the relationship between topography and function. It is also clear that an adequate understanding of these two issues can only be achieved through the use of a suitable technique for characterization of the topography. Such a characterization procedure involves both visual and numerical techniques.

Throughout the manufacturing organization, many groups require an understanding of the surface topography of materials and components if high quality products are to be consistently produced. For example, the designer requires information which will ensure that the surface characteristics specified are suitable for a known function, and must be able to define appropriate parameters and tolerance levels for satisfactory operation. The manufacturing engineer needs to understand the surface structural

requirements in order to specify a manufacturing process which will produce a surface having the necessary characteristics. The quality engineer needs to have a system which will provide sufficiently detailed information, relative to the specified characteristics, to ensure that surfaces have been produced within the pre-determined tolerance levels.

SURFACE CHARACTERIZATION

The most common method of determining surface characteristics is through the use of a stylus-based measuring instrument. The stylus is drawn across the surface at near constant velocity for a pre-determined distance. The vertical excursions of the stylus, relative to a datum, when suitably magnified, are a measure of the deviations of the real surface and within the calibrated tolerance range of the instrument.

In recent years metrologists and engineers have begun to realize that surfaces cannot be adequately characterized in two dimensions (2-D) and consequently emphasis has been given to three-dimensional (3-D) analysis. Since conventional stylus instruments are constrained to move in a straight line, a scanning technique may be implemented in order to accomplish 3-D assessment. The technique of recording or presenting data as a series of parallel traced lines is referred to as a 'raster' scan and can be achieved through the use of a stylus measuring instrument, employing a linear translation stage to move the work normal to the line of traverse.

As with conventional 2-D profilometry, a visual image of the logged area is desirable, if only to verify the system operation. More importantly, since 3-D quantitative analysis and modelling has only been investigated beyond its preliminary stages by some statisticians, qualitative analysis provides useful comparative information for the engineer. Thus, 2-D projections of the 3-D data, in the form of an axonometric plot and a contour map, accompanied by suitable numerical analysis are essential features of a surface characterization system.

THE 3-D DATA COLLECTION SYSTEM

The 3-D logging system used to obtain the data for the surfaces shown in this atlas is based on a modified Rank Taylor Hobson (RTH) Talysurf 5 surface measuring instrument incorporating a linear translation stage (Plate 1). In this system, area maps are logged by specifying sample spacing, trace spacing, number of data points per trace and number of traces. The specimen is moved normal to the line of traverse of the stylus by means of a linear translation stage. A raster scan of surface areas is achieved by indexing the stage a pre-programmed distance (trace spacing) before the start of each trace. The logged areas for each surface presented in Part I were square with sides measuring 1.304 mm. The grid spacing on both axes was 8μm, giving a total of 26 896 data points for each sample. The material used in all machining processes was a free cutting mild steel.

The datum plane in the direction of traverse can be accurately defined using a straight line datum attachment. Backlash problems have been eliminated by moving the work surface in only one direction for each profile. It should be noted that the stage is stationary during data logging, thus eliminating vibration problems and the need to accurately control the velocity of table movement.

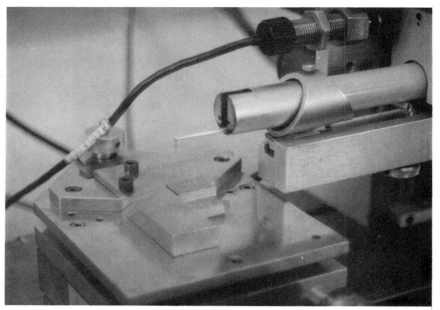

Plate 1

THE 3-D DATA ANALYSIS SYSTEM

When analysing data in three dimensions it is necessary to determine a suitable datum, such as that specified by the least squares mean plane. This plane is defined as the mean true plane, for which the sum of the squares of the errors yields a minimum value. When examining surfaces which are curved, such as those produced by turning, it is often beneficial to remove the curvature mathematically from the data. This process assists data manipulation and surface visualization. The best fitting curve which approximates to a given data set is that for which the sum of the residual squares is a minimum. For this work the least squares parabola was found to fit the data best and, where appropriate, this curve is subtracted from the original data prior to further manipulation.

As previously stated, 3-D data characterization is accomplished through the use of both visual and numerical techniques. In addition, specially developed numerical routines are used to manipulate the data for further analysis.

Numerical characterization

In this atlas numerical characterization involves the use of traditional 2-D parameters, their 3-D counterparts and volume-based measurements.

The most readily accepted measures of surface roughness characteristics are contained in the surface height amplitude distribution. Thus, when defining roughness parameters for an area, it is logical to consider the ordinate height distribution of the area as a measure of roughness.

Numerical integration techniques have been used to calculate the material volume of the logged area. By computing this volume before and after truncation (see below), two further parameters of tribological interest can be obtained – void volume and debris volume. The void volume parameter provides an estimation of the lubrication capacity of the surface in cubic mm per square mm of surface area. The debris volume provides an estimation of the expected volume of truncated material in the surface lubricant. The value is obtained as the volume difference between successive truncation levels in cubic mm per square mm of surface area.

To investigate directional properties of the surfaces we have presented graphs showing the distribution of single profile values in two orthogonal directions. The calculations of the profile parameters, Ra', and Rq', skew and kurtosis, are based upon the definitions given in ISO R468.

The following parameters are used in the atlas.

Area parameters
The following definitions are given for a set of data points Z_i where $1 \leqslant i \leqslant T$. T is the total number of data points (i.e., the product of the number of points per trace and the number of traces).

Mean = area mean height, $\bar{z}_a$

$$\bar{z}_a = \sum_{i=1}^{T} \frac{z_i}{T}$$

Ra = average roughness of area

$$Ra = \sum_{i=1}^{T} \left| \frac{z_i - \bar{z}_a}{T} \right|$$

Va = average variance

$$Va = \sum_{i=1}^{T} \frac{(z_i - \bar{z}_a)^2}{T}$$

Rq = root mean square roughness of area

$$Rq = \sqrt{Va}$$

Rsk = skewness of the area about the mean

$$Rsk = \sum_{i=1}^{T} \frac{(z_i - \bar{z}_a)^3}{TVa^{3/2}}$$

Rku = kurtosis of the area about the mean

$$Rku = \sum_{i=1}^{T} \frac{(z_i - \bar{z}_a)^4}{TVa^2}$$

Rt = the vertical height between the highest and lowest points of the area

Rp = distance from mean to highest peak

Rv = distance from mean to lowest valley

Volume parameters
Material volume (mm^3mm^{-2})

Void volume (mm^3mm^{-2})

Debris volume (mm^3mm^{-2})

Graphical characterization

Various graphical techniques have been incorporated to enhance visualization of the 3-D data. These include the following.

Axonometric plotting

This feature produces scaled 2-D projections of the 3-D data, including complete hidden line removal. The projections presented here use differential magnification to enhance visualization of asperity characteristics. This feature is consistent with 2-D analysis but the observer must always recognize the distortions which are inherent in the technique. Scale values on the z axis have the units of micrometers.

Contour plotting

This feature produces a surface map of equivalent height contours. The number of contour levels can be selected so as to produce a clear view of surface characteristics. The feature enhances qualitative analysis of surface details such as directionality and extends the use of surface topography data to include surface flaws or wear scars.

Data manipulation

In addition to the data visualization features of the package, the data can be manipulated to produce quantitative information, with particular reference to tribological functions and surface flaw analysis. The data manipulation features include the following.

Inversion

This provides a detailed view of the surface valleys in the inverted form. The feature is of particular use in revealing features in lubrication and wear analysis where the extent of the surface structure is important.

Truncation

The truncation feature is used both to predict controlled wear behaviour of surfaces in a tribological environment and to examine sub-surface features. The surface can be truncated to any pre-determined level and then analysed in any of the previously mentioned modes. The truncated data is used to compute a percentage area of contact parameter (contact %).

The truncation feature is used in combination with parameter calculation routines to allow the development of certain parameters to be analysed throughout the life (i.e., 1% to 100% truncation) of the surface. The results of this 'progressive truncation model' are presented in the form of composite graphs. For this atlas the development and interpretation of the following parameters are presented: contact %; material volume; void volume; debris volume; *Rsk* and *Rku*.

THE FIGURES

The figures presented in sections 1–14 of Part 1 of the atlas, are arranged in a standard sequence with some supplementary figures being added where appropriate.

Firstly, information is presented regarding the original unmodified surface in the form of an axonometric projection, height distribution with associated parameters and 5-level contour map. Height distribution information is presented as relative frequency (percentage of total data points) against height value (μm).

In order to investigate the directional properties of the surface and the variability of single trace parameters, four graphs are given which characterize trace values in two orthogonal directions. The relative frequency (percentage of total number of traces) of four parameters (*Ra'*, *Rq'*, skewness and kurtosis) are given.

In order to investigate the predicted behaviour of the surfaces in a tribological wear environment, they are subjected to a 30% and 70% truncation process. Figures included to indicate the effects of 30% truncation are an axonometric projection of the modified surface, a 5-level contour map, an axonometric projection of the debris removed, and the height distribution and associated area parameters for the modified surface.

To reveal the effects of 70% truncation the following figures are included: an axonometric projection of the modified surface, a 5-level contour map, an axonometric projection of the inverted surface to reveal the shape of the remaining valleys, and the height distribution and associated area parameters of the modified surface.

Finally, a list of figures is included to show the behaviour of surface parameters when subject to progressive truncation.

SUMMARY

The objective of the atlas is to provide a reference document of engineering surfaces so that the character of these surfaces may be understood in a 3-D sense.

Part 1 considers typical surfaces generated by 14 machining processes and describes the character of each in relation to the process by which it has been generated.

Part 2 provides an investigation that relates a machined surface's topographical features to its functional behaviour in a wear environment.

The Appendix provides information on relevant international standards for surface finish to provide the user with a reference source of useful literature.

PART 1

MACHINED SURFACES

1 TURNED SURFACES

The figures presented in this section relate to a surface typical of those produced by the turning process. In this case, for ease of instrumentation, a face-turned surface has been selected, but the characteristics would be similar for any turned surface. Figures 1.1 and 1.3 show that such a surface has a highly regular structure generated by the single point of the cutting tool moving across the surface at a constant feed-rate during machining. Note that the structure generated is slightly curved, the rate of curvature being dependent upon the distance of the tool point from the centre of the workpiece. The structure of the surface is therefore a combination of a waviness component, generated by the feed process associated with the cutting point, and a random component of micro-roughness, caused by the action associated with chip removal. The randomness is due to the built-up edge on the cutting tool being continuously generated and removed during machining; the rate at which this occurs is partly determined by non-homogeneity within the body of the material itself.

As is evident from Figure 1.15, in a tribological situation this surface will yield rapid asperity removal until approximately 50% of the original Rt value has been lost; after this stage has been reached the rate of debris removal becomes more or less constant. In general, the turned surface appears to have potential as a tribological ssurface since the valleys provide the lubricant retention features needed in any surface. To improve the surface characteristics it is essential to pay considerable attention to the shape of the tip of the surface-generating cutting tool.

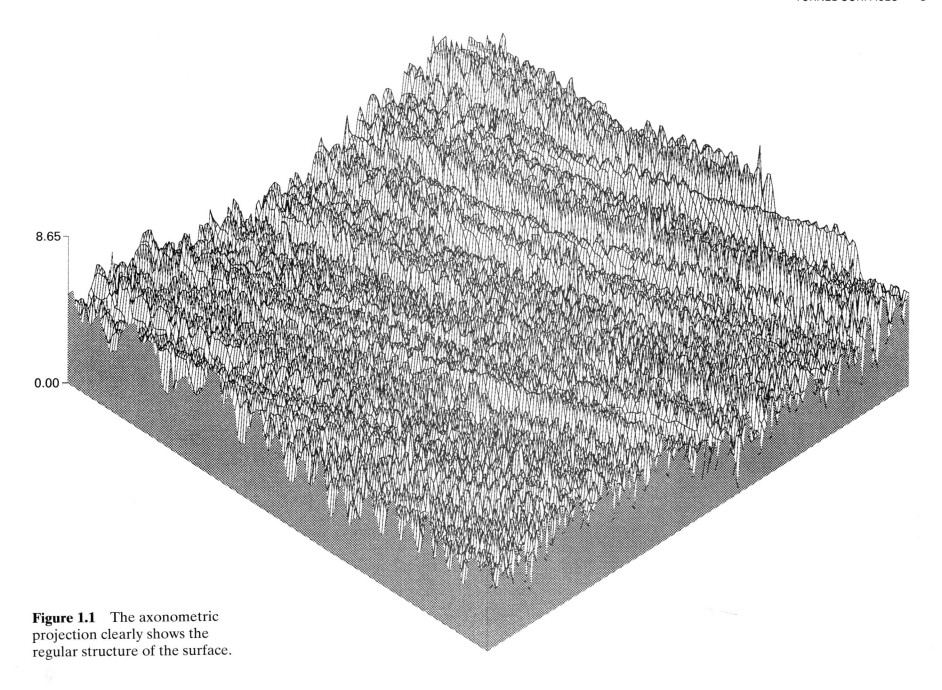

Figure 1.1 The axonometric projection clearly shows the regular structure of the surface.

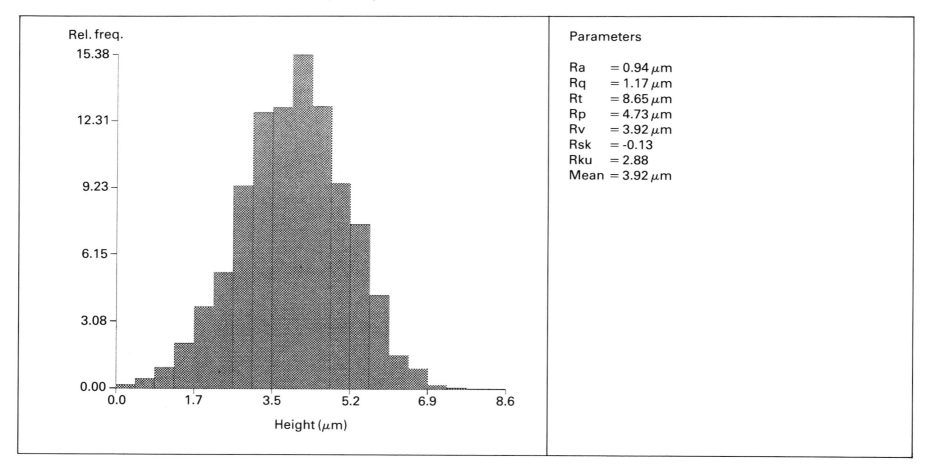

Rel. freq.

Parameters

Ra	= 0.94 μm
Rq	= 1.17 μm
Rt	= 8.65 μm
Rp	= 4.73 μm
Rv	= 3.92 μm
Rsk	= -0.13
Rku	= 2.88
Mean	= 3.92 μm

Height (μm)

Figure 1.2   This histogram shows that the distribution of the asperities is closely Gaussian, indicated by values of the 3-D skewness and kurtosis ($Rsk = -0.13$; $Rku = 2.88$). These values are typical of those found in a turning process, whether it has been produced by face turning or cylindrical turning. The magnitude of the Ra parameter can vary dramatically during face turning as it is dependent on the shape of the cutting edge, the depth of the cut and the feed-per-revolution of the cutting tool.

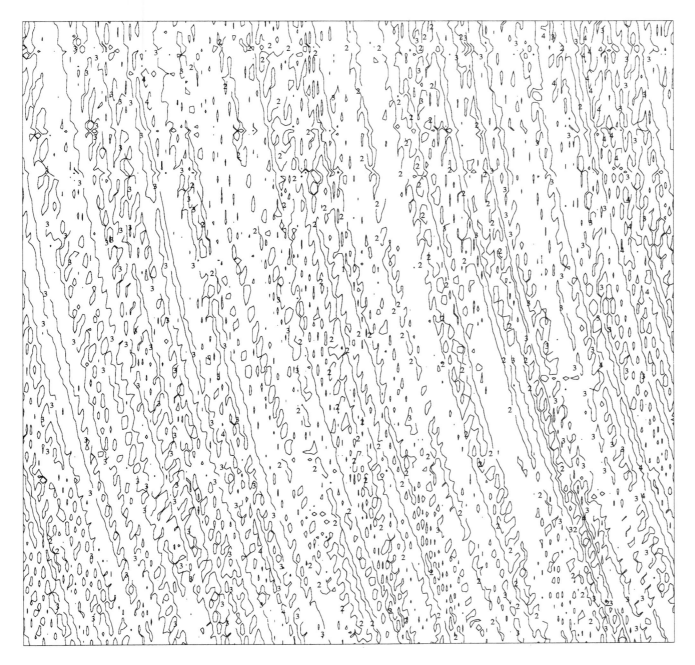

Contour key (μm)

1 : 0.86   3 : 4.32   5 : 7.78
2 : 2.59   4 : 6.05

Figure 1.3   The contour map shows the parallel nature of the cutting process, although it is important to note that the surface generated exhibits a curved structure which is a distinctive feature of face turning. The slight deviations are due to micro-disturbances caused by the cutting mechanics discussed in the text.

Figure 1.4 (over)   The information on 2-D profile parameters presented here has been extracted from the 3-D surface data shown in Figure 1.1. The turned surface, as indicated earlier, exhibits significant directional properties, as shown by the individual distributions compiled from the 2-D profile data. The first two graphs, relating to asperity heights in terms of Ra' and Rq' indicate that the average height of the profiles is significantly greater when measured along an axis running across the direction of tool feed. This indicates that the surface is highly anisotropic. The distribution of surface shape parameters, skewness and kurtosis, also exhibit differences, but in these cases the discrepancy is much smaller.

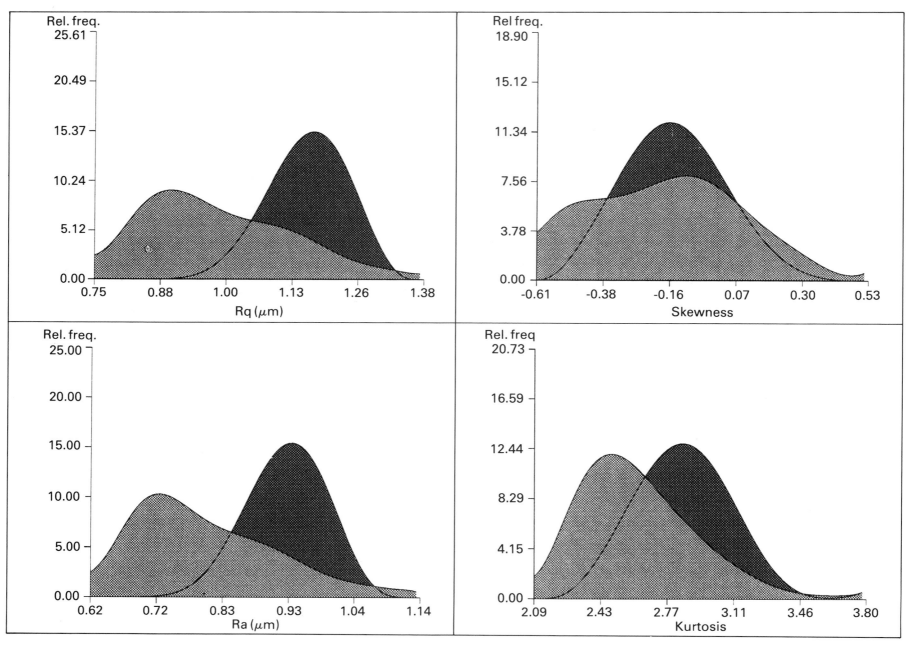

Figure 1.4

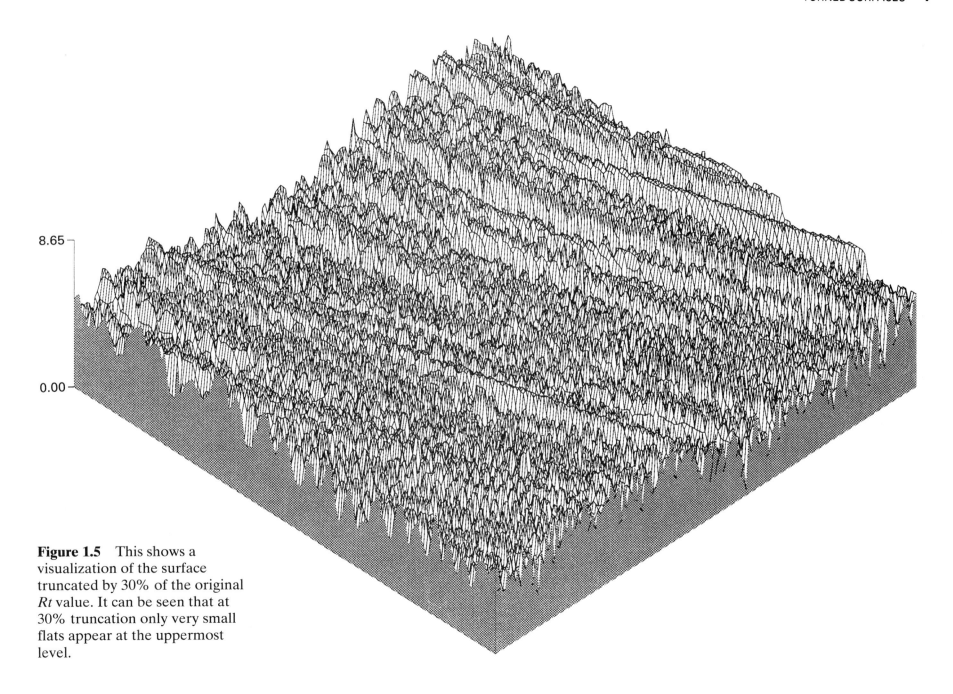

8.65

0.00

Figure 1.5 This shows a visualization of the surface truncated by 30% of the original *Rt* value. It can be seen that at 30% truncation only very small flats appear at the uppermost level.

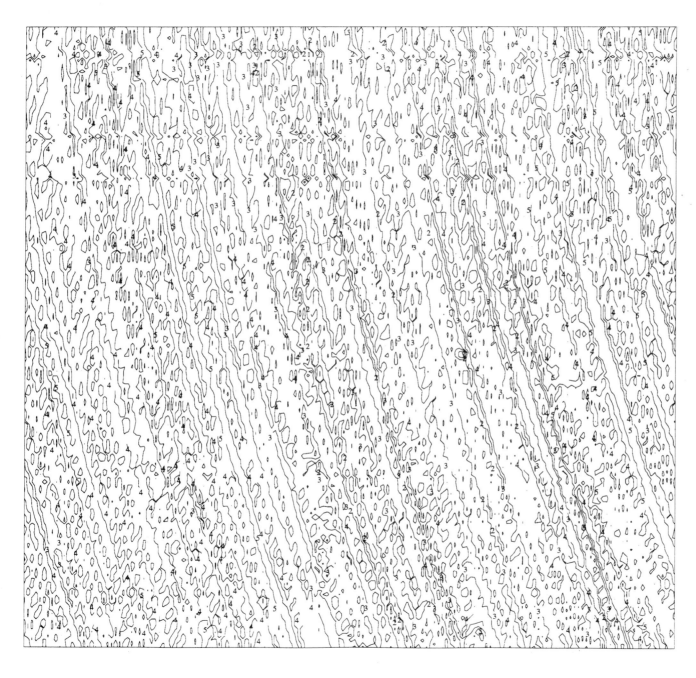

Contour key (μm)

1 : 0.60
2 : 1.81
3 : 3.02
4 : 4.23
5 : 5.44

Figure 1.6 This contour map is similar to, if not indistinguishable from, the previous map shown in Figure 1.3, indicating that large structural changes have not yet occurred.

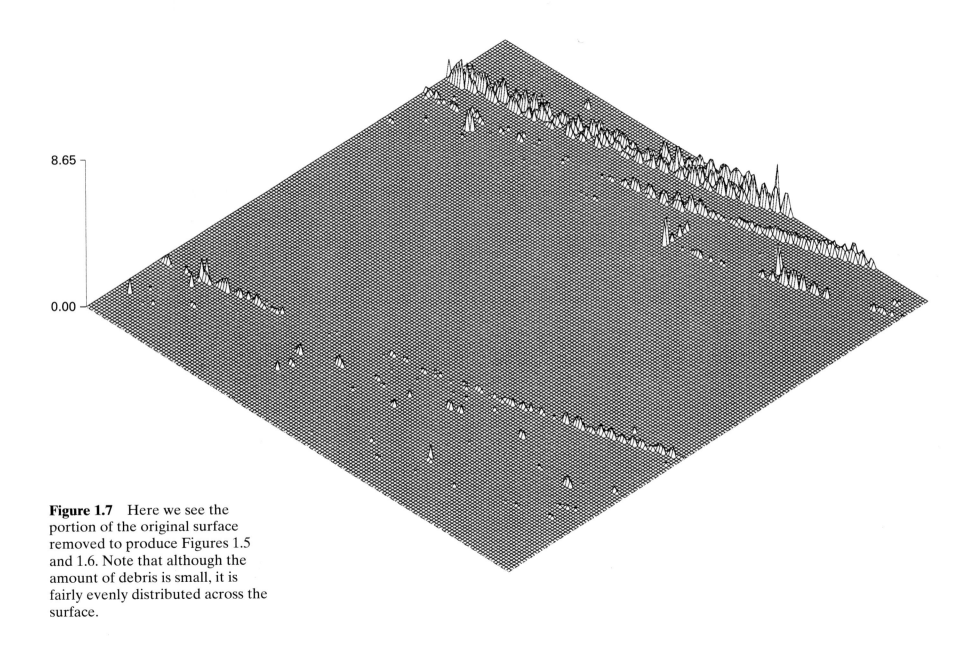

8.65

0.00

Figure 1.7 Here we see the portion of the original surface removed to produce Figures 1.5 and 1.6. Note that although the amount of debris is small, it is fairly evenly distributed across the surface.

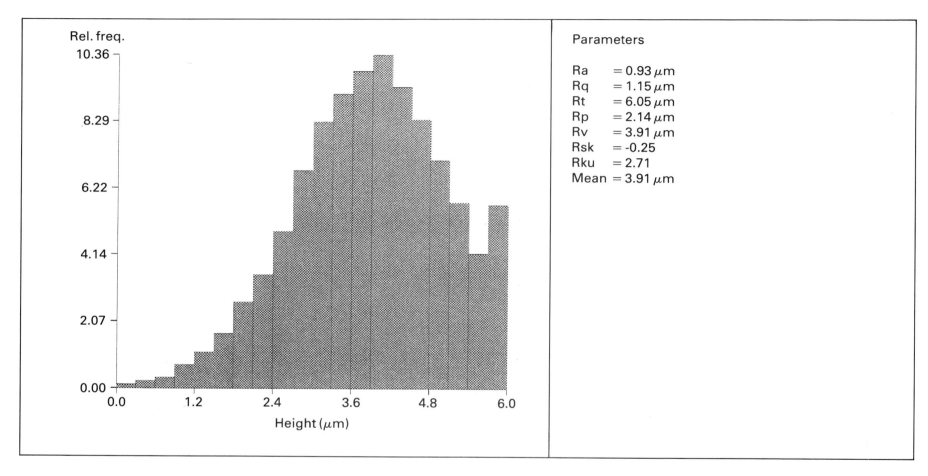

Parameters

Ra   = 0.93 μm
Rq   = 1.15 μm
Rt   = 6.05 μm
Rp   = 2.14 μm
Rv   = 3.91 μm
Rsk = -0.25
Rku = 2.71
Mean = 3.91 μm

Figure 1.8 Truncation of the surface by 30% *Rt* does not significantly affect the height parameters, *Ra* and *Rq*. However, it can be seen that the surface has become more negatively skewed, and that the value of kurtosis has decreased from its original value; a feature of the development of the kurtosis parameter as it changes with truncation.

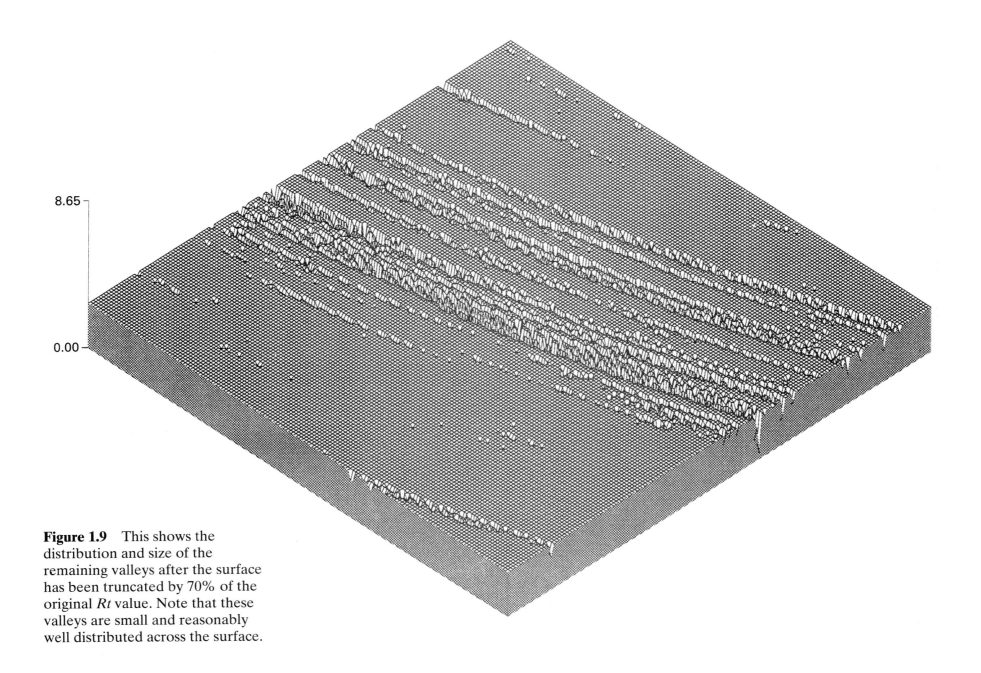

8.65 —

0.00 —

Figure 1.9   This shows the
distribution and size of the
remaining valleys after the surface
has been truncated by 70% of the
original *Rt* value. Note that these
valleys are small and reasonably
well distributed across the surface.

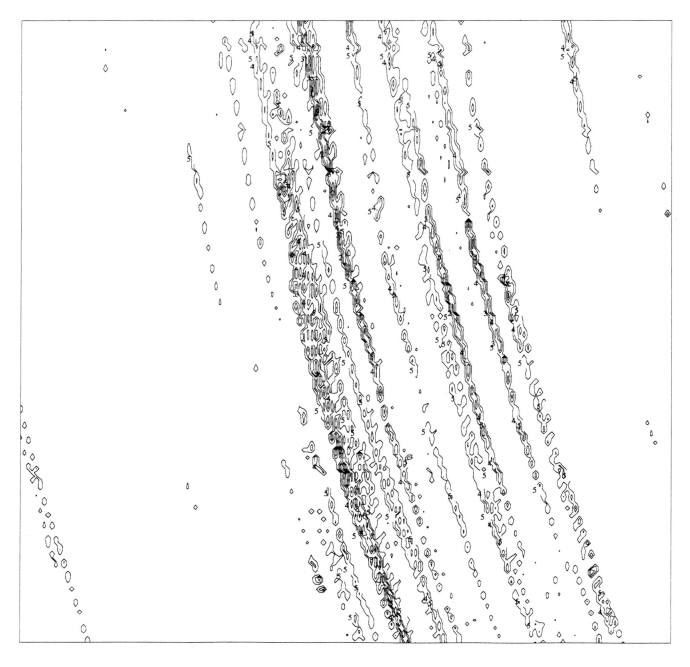

Contour key (μm)

1 : 0.26
2 : 0.78
3 : 1.30
4 : 1.82
5 : 2.34

Figure 1.10 After truncation by 70% *Rt* the contour map still indicates the curved nature of the remaining valleys, which were produced as a consequence of the original machining process.

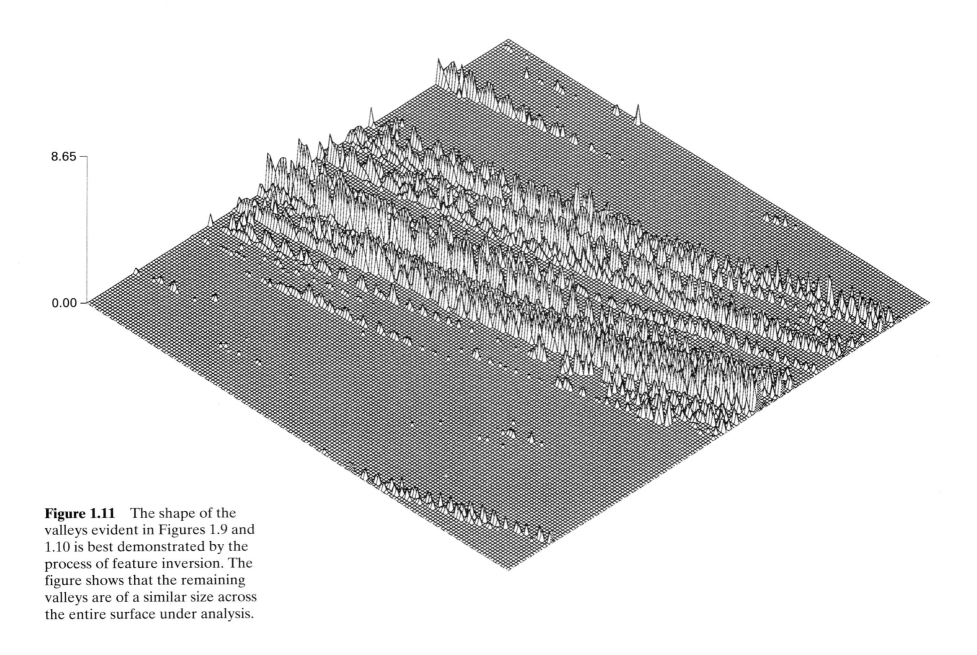

8.65

0.00

Figure 1.11 The shape of the valleys evident in Figures 1.9 and 1.10 is best demonstrated by the process of feature inversion. The figure shows that the remaining valleys are of a similar size across the entire surface under analysis.

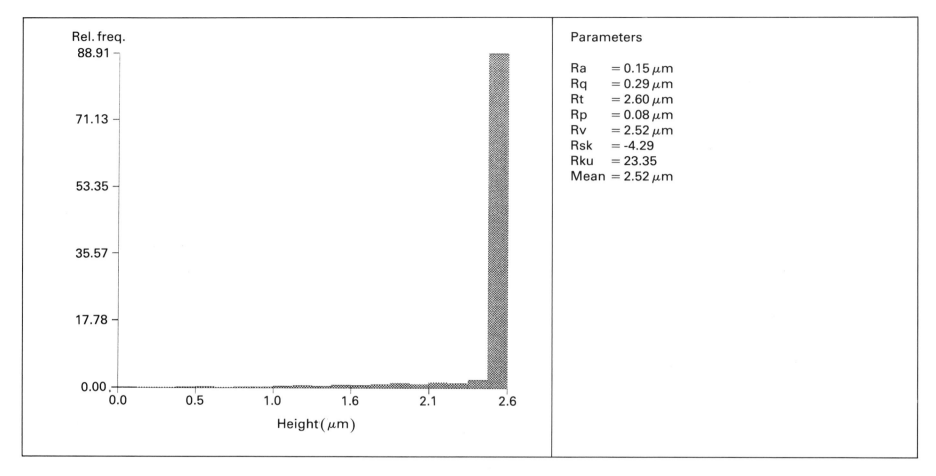

Figure 1.12 The height distribution and associated parameters of the surface after 70% *Rt* truncation shows the proportion of the surface at the uppermost levels, leading to a very high negative skewness and high kurtosis values (*Rsk* = –4.29; *Rku* = 23.35).

Figure 1.13 (over) This graph illustrates how the value of contact % varies as the original surface is truncated in steps of 1% *Rt* (simulated wear). The nature of the curve on the graph indicates that there is a slow transition from low contact % to high contact % as the surface is truncated, which may be a very useful feature in a tribological surface.

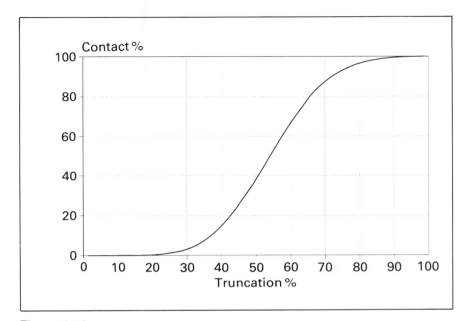

Figure 1.13

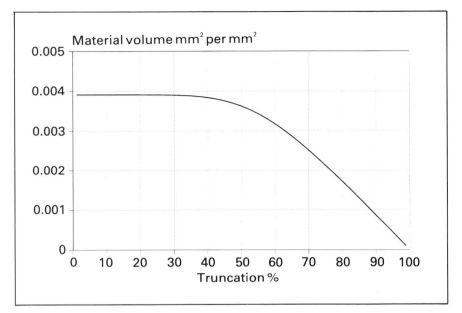

Figure 1.14

Figure 1.14   Here data on material volume remaining is plotted against truncation %. It can be clearly seen that the rate of reduction of material is relatively constant after the 70% *Rt* truncation level has been reached. This indicates that although the volume of the valleys does not remain constant with truncation, the mass of body material which is being removed is obscuring the valley information.

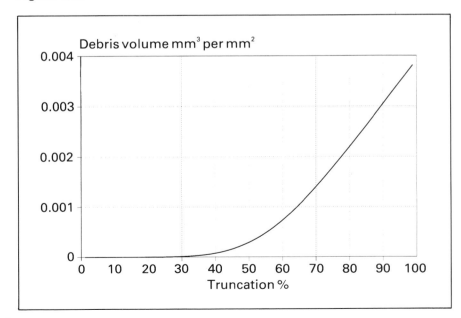

Figure 1.15

Figure 1.15   Here debris volume is plotted against truncation %, another useful method of analysis. As would be expected from the earlier figures, the debris volume generated is very small at the early stages of truncation. The rate of debris removal becomes approximately constant from the 50% level point onwards throughout the remainder of the truncation process. The non-linearities are again being caused by the bulk of material which is being truncated obscuring the information relating to the remaining valleys.

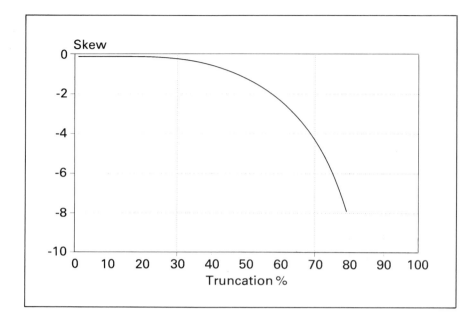

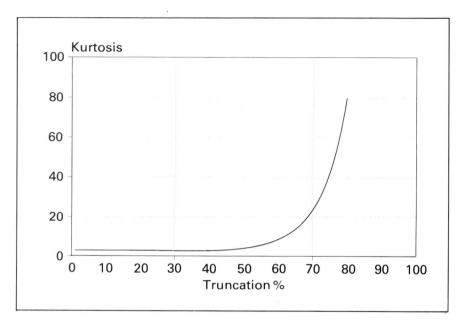

Figure 1.16 In this graph the area skewness parameter plotted is against truncation %. It can be seen that the skewness develops from an initial value of $Rsk = -0.13$ and decreases gradually, reaching a value of –8.0 at 80% truncation. Experience has shown that most tribological surfaces start to 'fail' by scuffing when the value of skewness is approximately –4.0, hence this surface preserves sufficient oil retaining features up to approximately 70% Rt truncation level. Re-examination of Figure 1.10 in this context shows that the oil retention grooves are well distributed and hence increased lubrication potential.

Figure 1.17 The kurtosis parameter is intimately related to skewness, and here it can be seen that its value remains substantially constant up to 50% Rt truncation. At this point kurtosis begins to increase rapidly until it reaches a value of $Rku = 23$ at 70% truncation.

2

ELECTRO DISCHARGE MACHINED SURFACE

The figures presented in this section relate to a surface typical of those produced by the electro discharge machining (or spark machining) process undertaken at a bulk machining setting. The surface has an extremely random structure, caused by the electrical discharges randomly impinging on the surface through the di-electric fluid during the machining process. As both the distribution and the intensity of the discharge are affected by the suspension of particles in the di-electric fluid, the cavities formed on the surface have differing magnitudes. The high peaks on the surface are formed adjacent to the melted valleys.

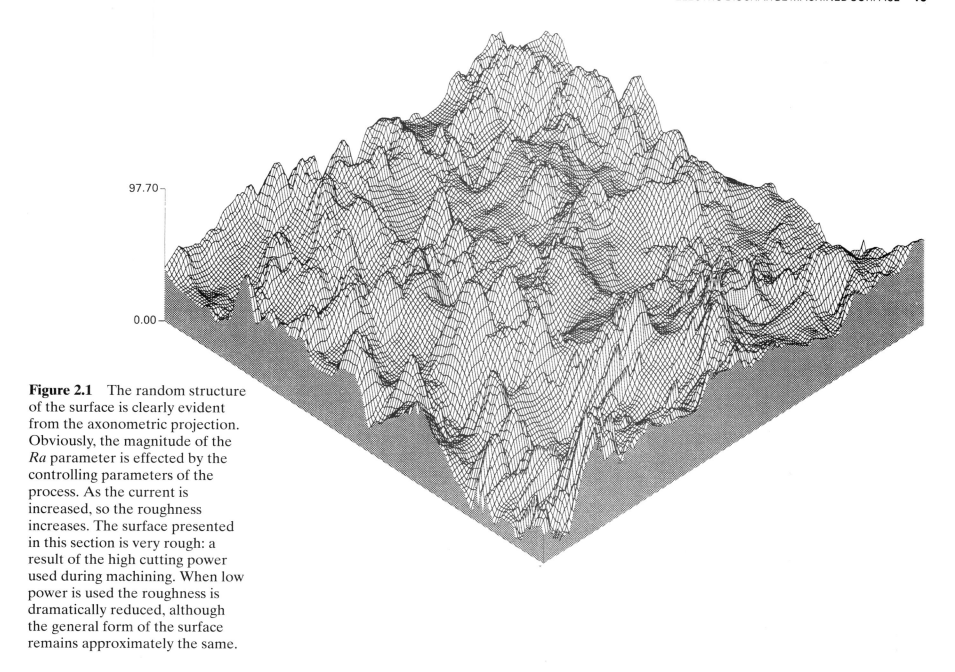

97.70 —

0.00 —

Figure 2.1 The random structure of the surface is clearly evident from the axonometric projection. Obviously, the magnitude of the *Ra* parameter is effected by the controlling parameters of the process. As the current is increased, so the roughness increases. The surface presented in this section is very rough: a result of the high cutting power used during machining. When low power is used the roughness is dramatically reduced, although the general form of the surface remains approximately the same.

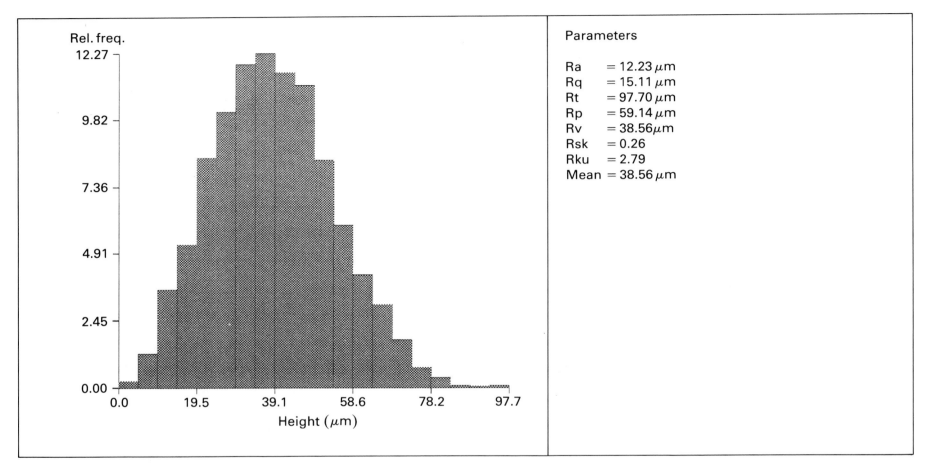

Figure 2.2 The histogram shows that the distribution of asperities on the surface is pseudo-Gaussian, as indicated by the 3-D skewness and kurtosis values ($Rsk = 0.26$; $Rku = 2.79$). Quite clearly, although no obvious correlation is present, a surface pattern is distinguishable. This is caused by the well-known phenomenon that sparking will always take the shortest route to the surface, eroding the highest peaks. In general, spark eroded surfaces do not exhibit directional properties, but such properties can be introduced, and subsequently determined, as a result of the influence of di-electric flow between the tool (spark source) and the component.

The fact that the voids disappear at a linear rate when the lower levels of the surface are truncated (see Figure 1.14) indicates that the energy in the spark decreases linearly at lower surface levels as the spark's kinetic energy is dissipated into the bulk material.

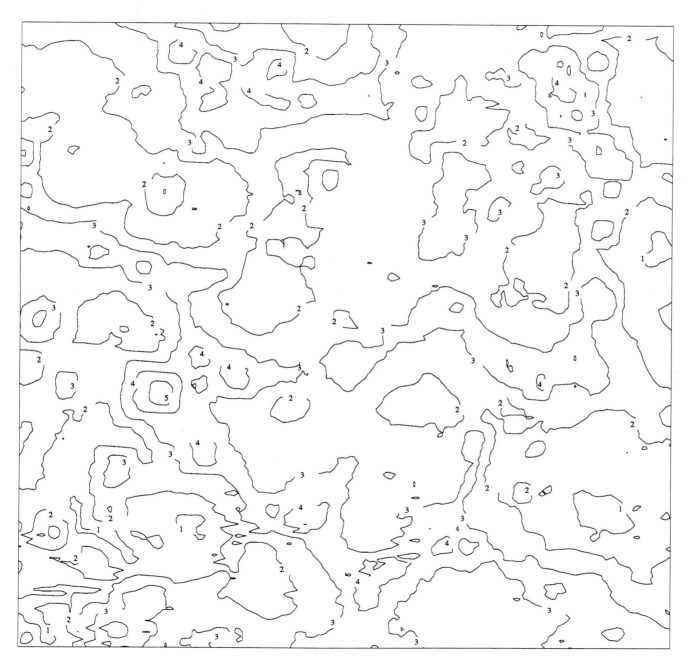

Contour key (μm)

1 : 9.77	4 : 68.39
2 : 29.31	5 : 87.93
3 : 48.85	

Figure 2.3 The contour map shows five levels of contours. Again, the figure indicates that the machining process produces a surface with a random structure.

Figure 2.4 (over) The information presented here relates to 2-D profile parameters extracted from the surface in two orthogonal directions. The electro-discharge process is expected to produce a random isotropic surface. The four sets of distribution curves indicate a reasonably close similarity between two orthogonal directions of assessment. The distribution of the height parameters (Ra' and Rq') yield closely similar graphs although, as expected, the average magnitude of Rq' is greater than Ra' (the relationship is approximately $0.8 \, Rq' = Ra'$ which holds true for a genuinely Gaussian distribution of asperities). Very close similarities also exist in the shape parameters, skewness and kurtosis. The combined results confirm a genuinely random surface.

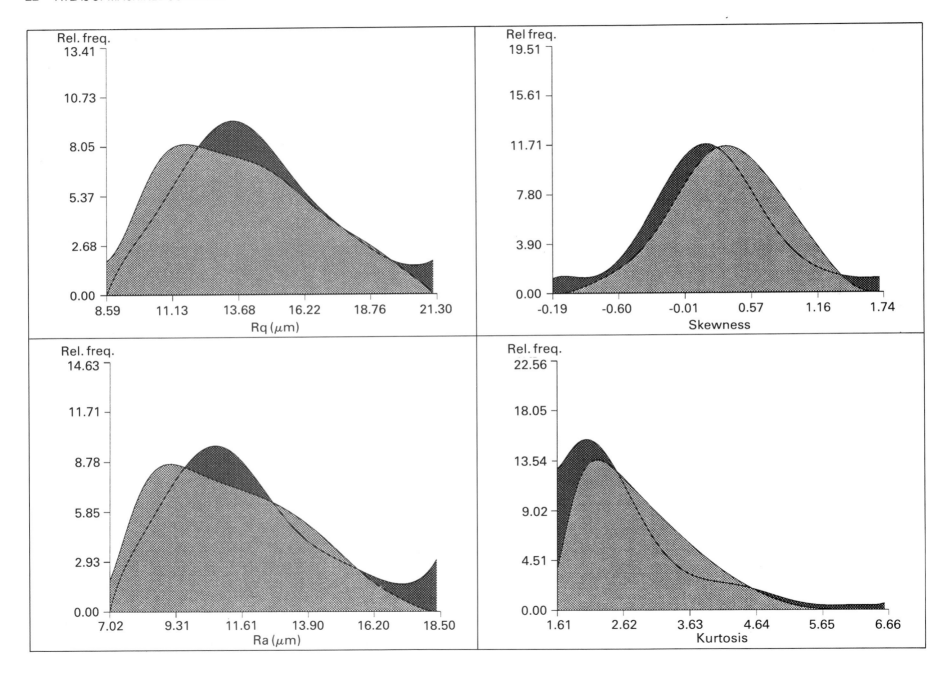

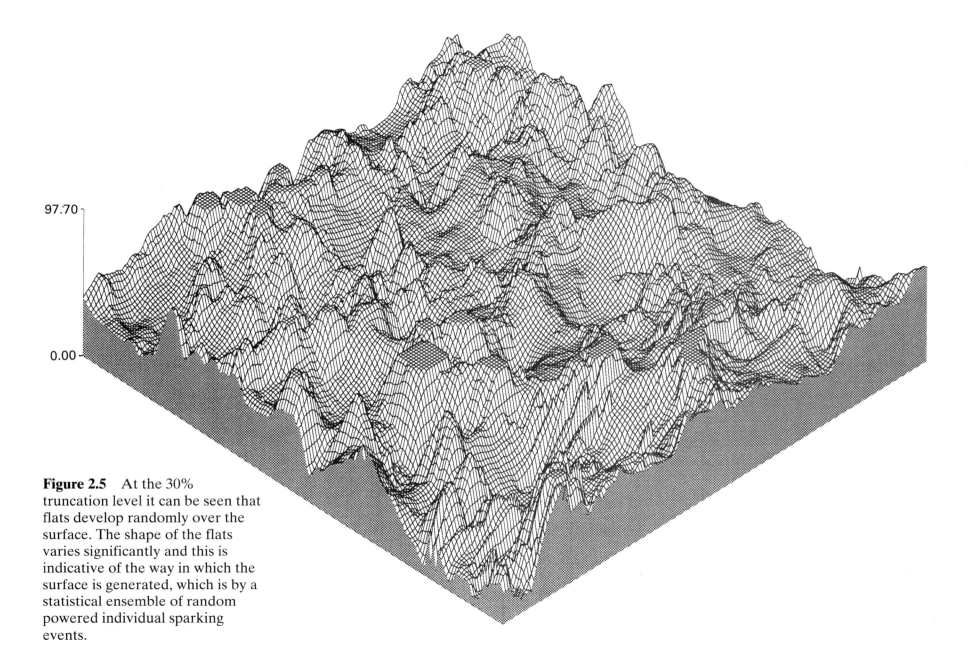

97.70

0.00

Figure 2.5 At the 30% truncation level it can be seen that flats develop randomly over the surface. The shape of the flats varies significantly and this is indicative of the way in which the surface is generated, which is by a statistical ensemble of random powered individual sparking events.

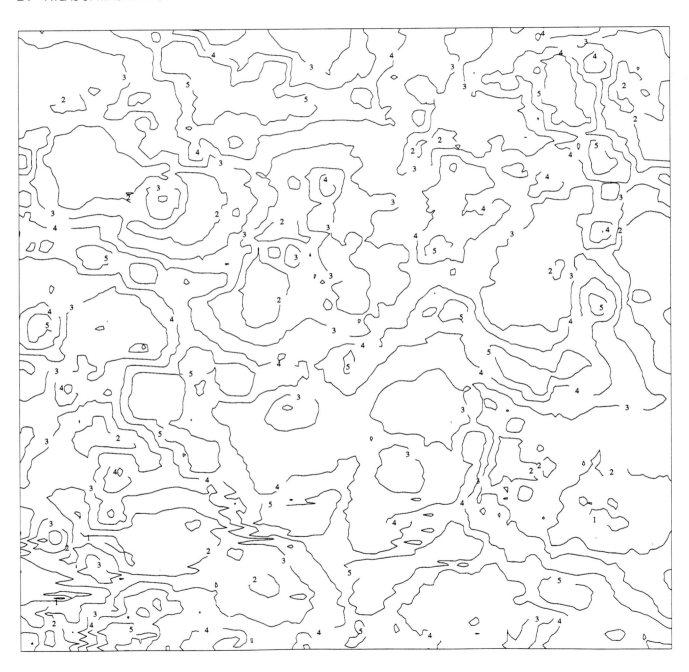

Contour key (μm)

1 : 6.84
2 : 20.52
3 : 34.20
4 : 47.88
5 : 61.56

Figure 2.6 When the contour map of the surface after 30% *Rt* truncation is compared to Figure 2.3 it can be seen that the general pattern of the surface is retained.

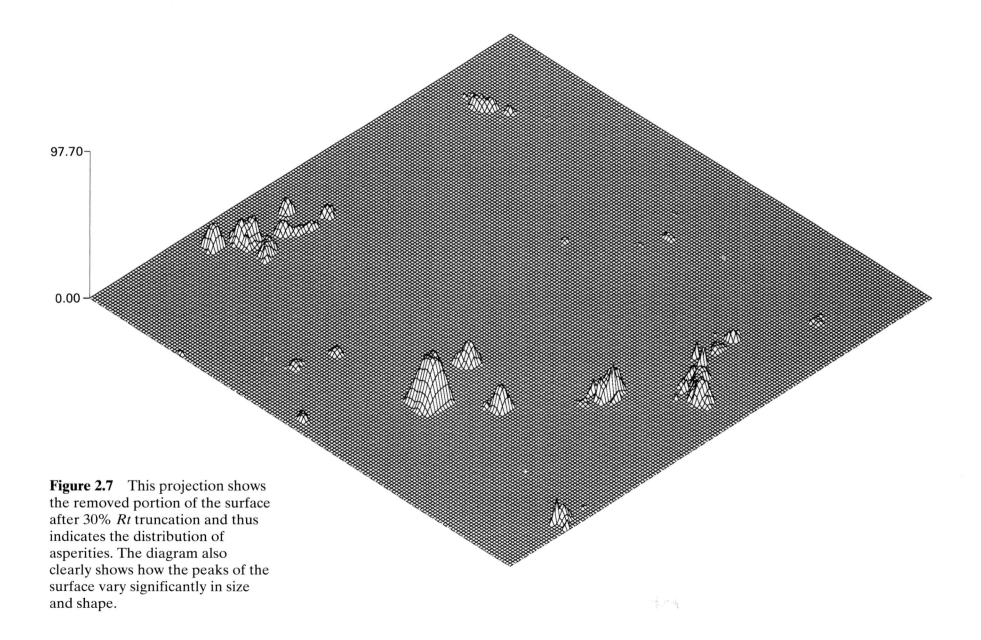

97.70

0.00

Figure 2.7 This projection shows the removed portion of the surface after 30% *Rt* truncation and thus indicates the distribution of asperities. The diagram also clearly shows how the peaks of the surface vary significantly in size and shape.

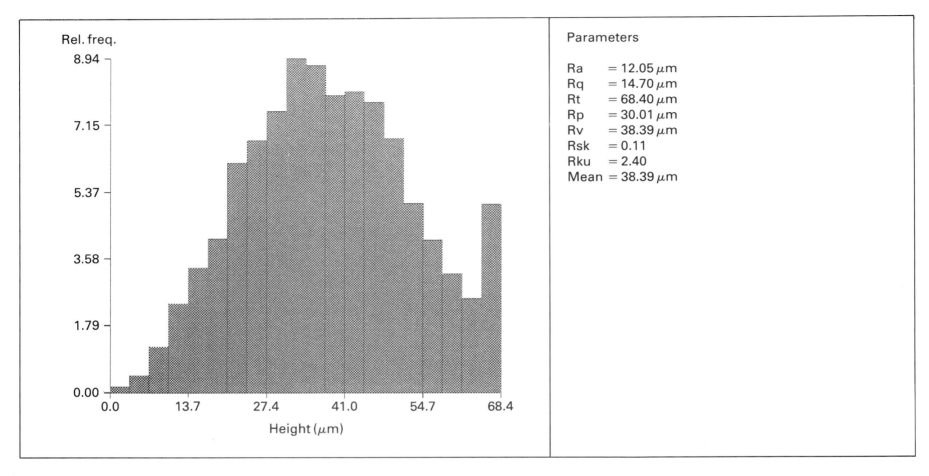

Figure 2.8 The effect on the height parameters of removing the upper 30% *Rt* is demonstrated in this histogram. The average roughness parameter has been reduced from 12.23 μm to 12.05 μm, a marginal change; *Rq* is similarly effected. The surface shape parameters, *Rsk* and *Rku*, have only marginally changed as a consequence of truncation, due to the small number of high peaks.

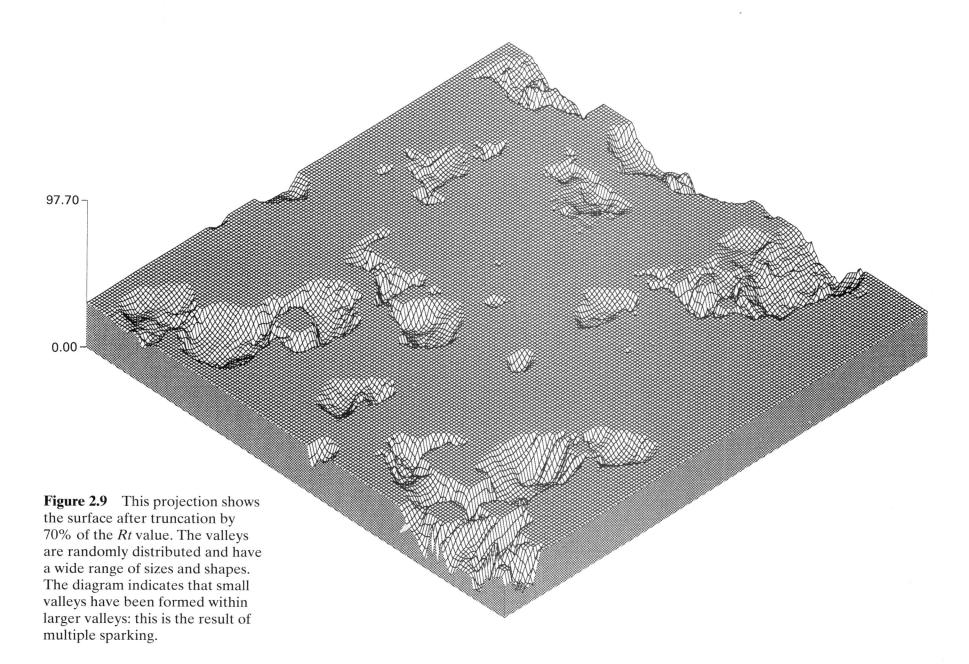

97.70

0.00

Figure 2.9 This projection shows the surface after truncation by 70% of the *Rt* value. The valleys are randomly distributed and have a wide range of sizes and shapes. The diagram indicates that small valleys have been formed within larger valleys: this is the result of multiple sparking.

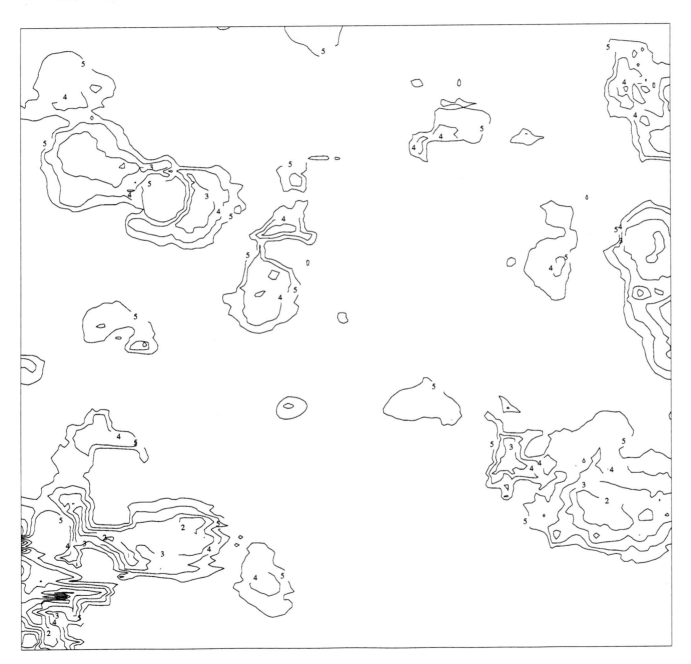

Contour key (μm)

1 : 2.93
2 : 8.79
3 : 14.65
4 : 20.51
5 : 26.37

Figure 2.10 This contour map is of the 70% truncated surface shown in Figure 2.9.

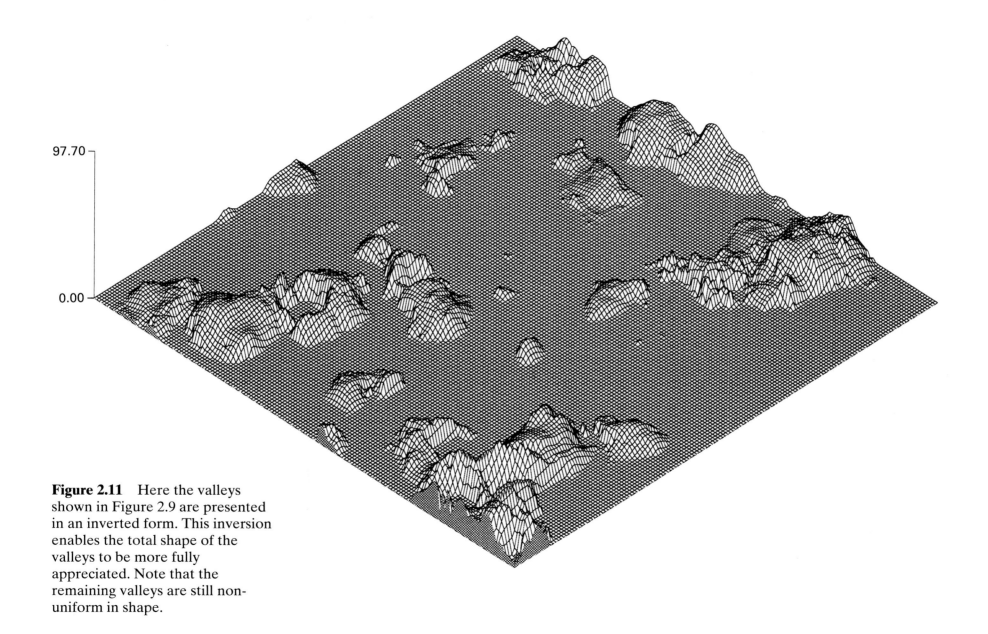

97.70

0.00

Figure 2.11   Here the valleys shown in Figure 2.9 are presented in an inverted form. This inversion enables the total shape of the valleys to be more fully appreciated. Note that the remaining valleys are still non-uniform in shape.

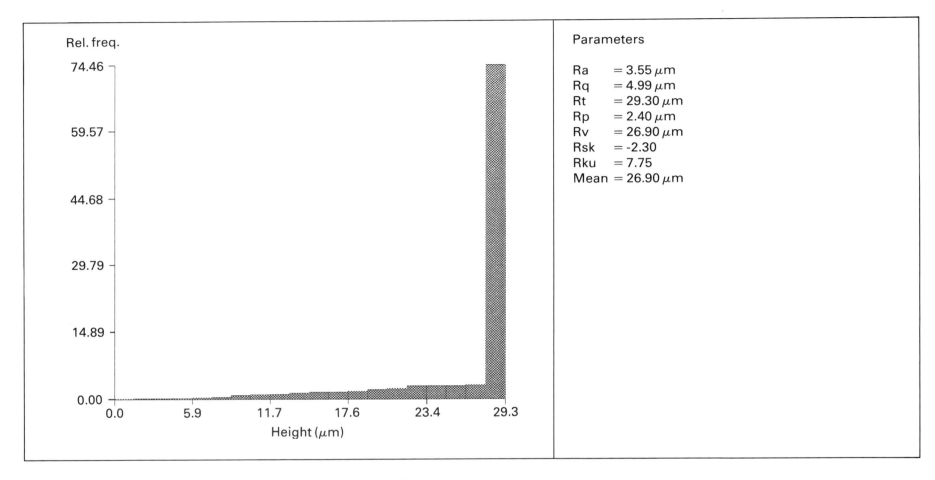

Figure 2.12 This figure presents the height distribution and associated parameters of the surface after 70% *Rt* truncation.

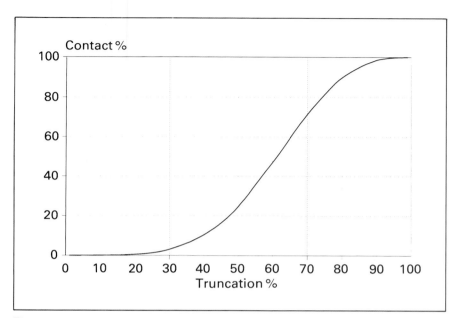

Figure 2.13

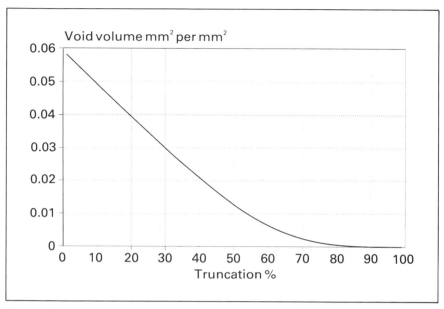

Figure 2.14

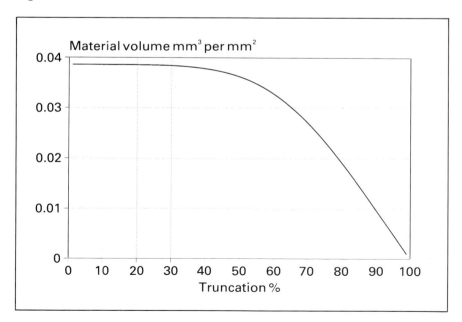

Figure 2.15

Figure 2.13 Here contact % is plotted against truncation %. The nature of the curve shows a slow but continuous transition from low contact % to high contact % as the surface is truncated. It can be seen that 50% contact will occur at a truncation level of approximately 62% Rt. Note that a very high percentage contact is achieved at 80% Rt truncation.

Figure 2.14 When the void volume parameter is plotted the graph shows that the reduction in void volume is generally linear up to approximately 50% Rt truncation and that the void volume reduces to 10% at 62% Rt truncation.

Figure 2.15 Here material volume remaining is plotted against truncation % showing that during the initial stages of truncation the rate at which material volume is removed shows little variation. At 40% Rt truncation the volume removed starts to steadily increase, reaching a maximum rate at approximately 70% Rt truncation. From 70% to 100% Rt truncation the slope of the graph becomes constant.

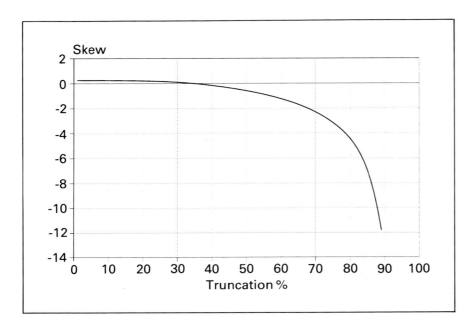

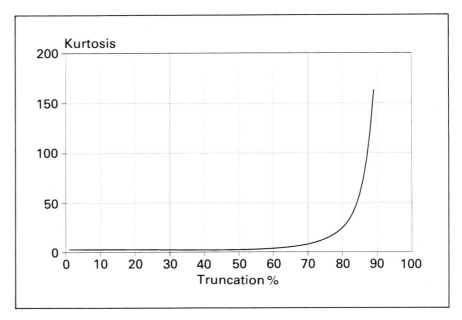

Figure 2.16 This graph indicates that skewness is hardly effected at all by truncation until 40% *Rt* truncation is reached. The initial value of skewness of this surface is marginally positive, and this is typical of most surfaces produced by the electro-discharge process.

Figure 2.17 The kurtosis value initially yields 2.79 (Figure 2.2) and remains constant until a truncation level of 50% *Rt*.

3 END MILLED SURFACES

The figures presented in this section relate to a surface typical of those generated by the end milling process. This process produces a surface with a highly correlated structure, generated by the rotation of the end mill as it is traversed across the surface of the material. The structure of the surface is related to the variation in surface cutting speed and the feed-rate as the cutter is traversed across the surface. The surface cutting speed is a product of the milling machine feed-rate and the end mill tool-face cutting velocity, which varies along the edge as the radius of the tool increases from the centre to the outside. The surface cutting speed approaches zero at the centre of the tool. As the end mill is traversed across the surface of the component cutting is occurring all along the cutting edge, and hence the cutting velocity, normally termed V (feet/minute or metres/second), varies significantly.

As the cutting velocity varies, the energy in the tool-chip interface also varies and a range of differing cutting mechanisms exist, including surface fatigue, plastic deformation and shear. As a consequence of these different cutting mechanisms, a wide range of asperity formations occur which contribute to primary roughness.

The magnitude of the Ra parameter can vary dramatically during end milling. It is dependent upon the tool rotation speed, the table traverse rate, the depth of cut and the condition of the cutting edge. But the general structure of the surface will remain substantially similar irrespective of the selection of these conditions. In a tribological sense the surface very rapidly generates good bearing-area properties and continues to retain a steady increase until 70% truncation, at which point the surface is becoming dominated by flat areas (see Figure 3.13). However, this surface has limited usefulness in a tribological environment since the oil-retaining capability is significantly reduced at relatively modest levels of truncation (see Figure 3.14).

The near Gaussian distribution of the asperity heights on the endmilled surface (see Figure 3.2) is typical of the surfaces produced by many machining processes.

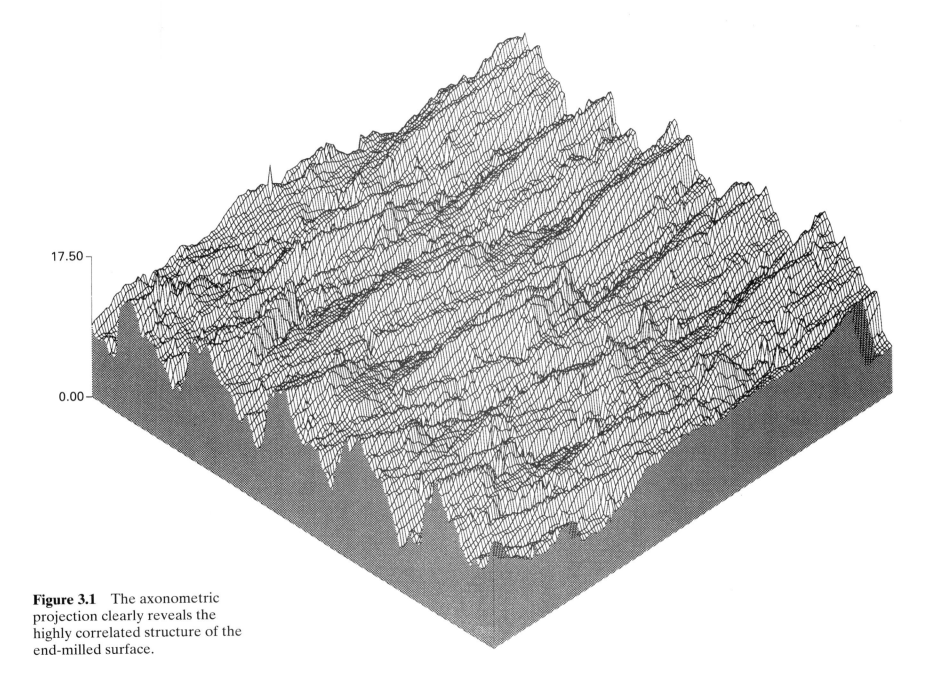

17.50 —

0.00 —

Figure 3.1 The axonometric projection clearly reveals the highly correlated structure of the end-milled surface.

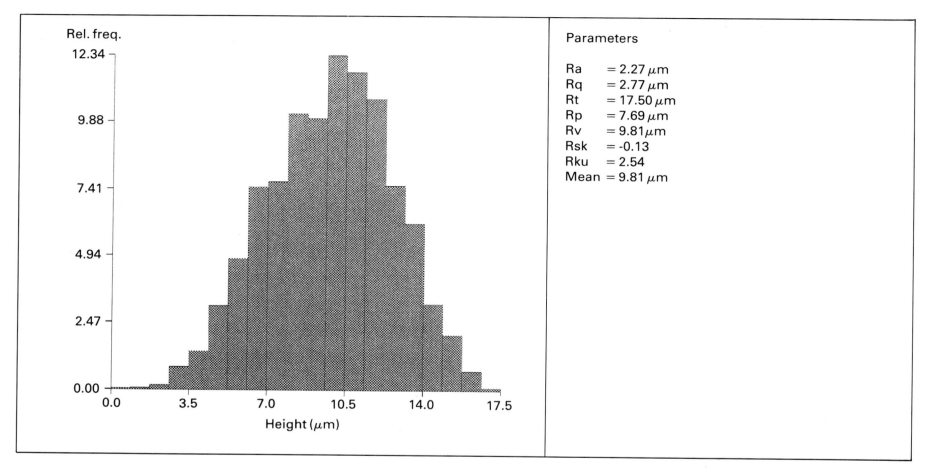

Figure 3.2 The histogram indicates that the distribution of the asperity heights of the end milled surface is closely Gaussian. This is demonstrated not only visually but also by the shape parameters of the surface (*Rsk* = –0.13; *Rku* = 2.54).

Contour key (μm)

1 : 1.75
2 : 5.25
3 : 8.75
4 : 12.25
5 : 15.75

Figure 3.3 The contour map shows the parallel nature of the cutting process. The plot also indicates that the surface has been logged in a direction which is not exactly in line with the machining process. Such accuracy is less important in 3-D visualization and characterization, but may be significant in 2-D assessment.

Figure 3.4 (over) The four sets of curves presented here show height information for 2-D profiles extracted from two orthogonal directions. The curves relate to the parameters Ra', Rq', skewness and kurtosis. Note that, in all four cases, the distribution of data extracted from the two profiles is significantly different, confirming the directional nature of the surface. This is not at all surprising for Ra' and Rq', but the extent of variation between the two distribution curves for the shape parameters, skewness and kurtosis is highly significant.

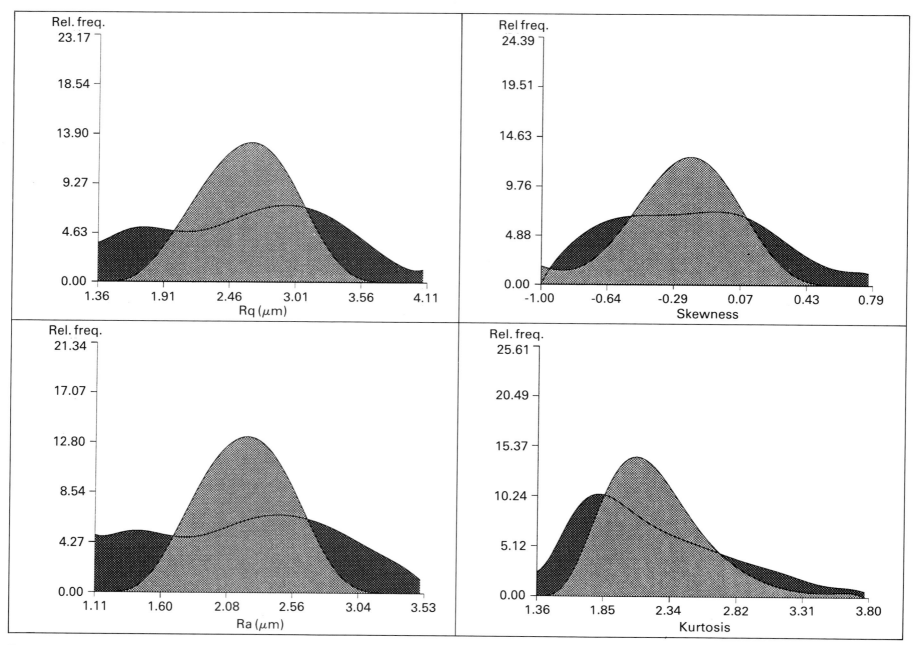

Figure 3.4

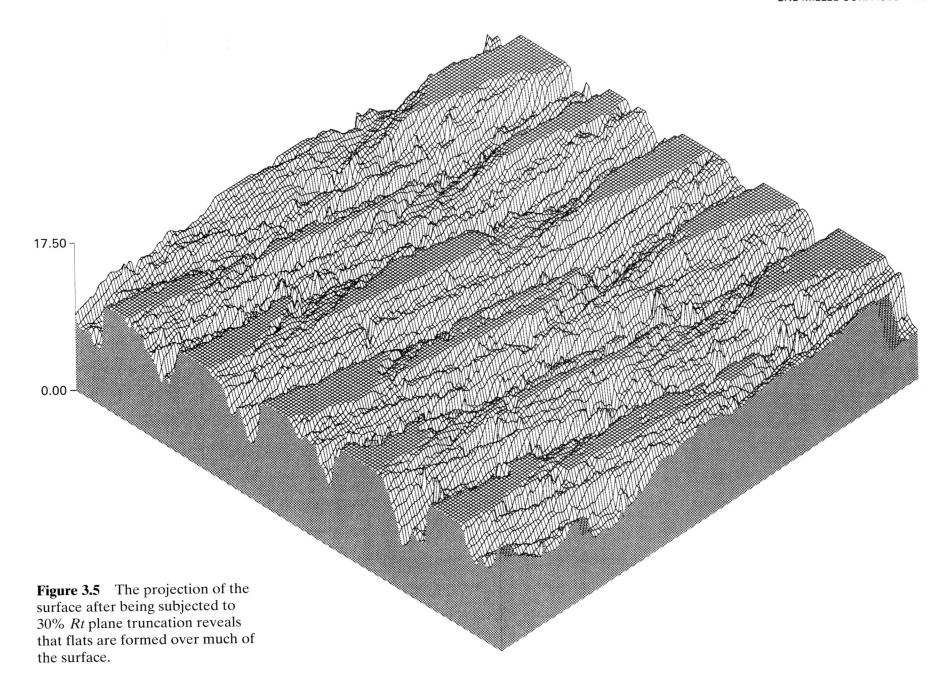

17.50 —

0.00 —

Figure 3.5 The projection of the surface after being subjected to 30% *Rt* plane truncation reveals that flats are formed over much of the surface.

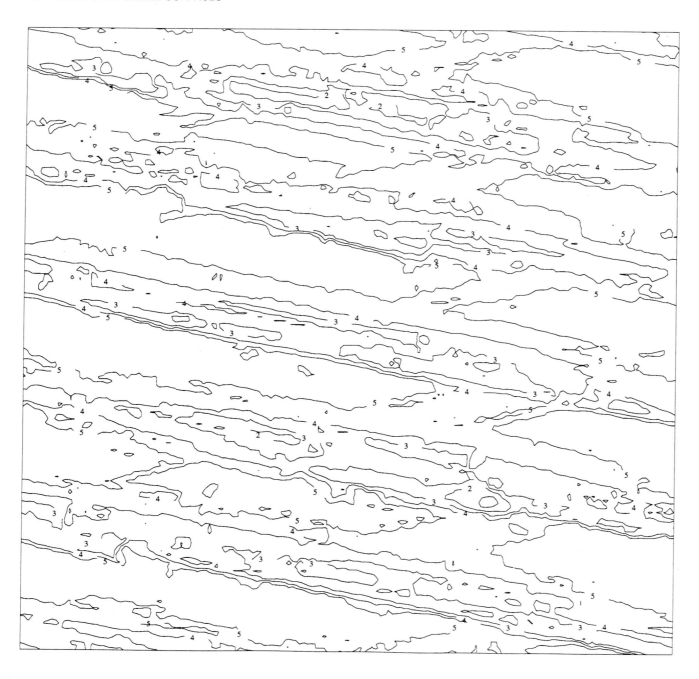

Contour key (μm)

1 : 1.22
2 : 3.66
3 : 6.10
4 : 8.54
5 : 10.98

Figure 3.6 The contour map of the surface after 30% *Rt* truncation shows that the area of the flats evident in Figure 3.5 varies significantly; those at the edges of the investigated surface being wider than those at the centre.

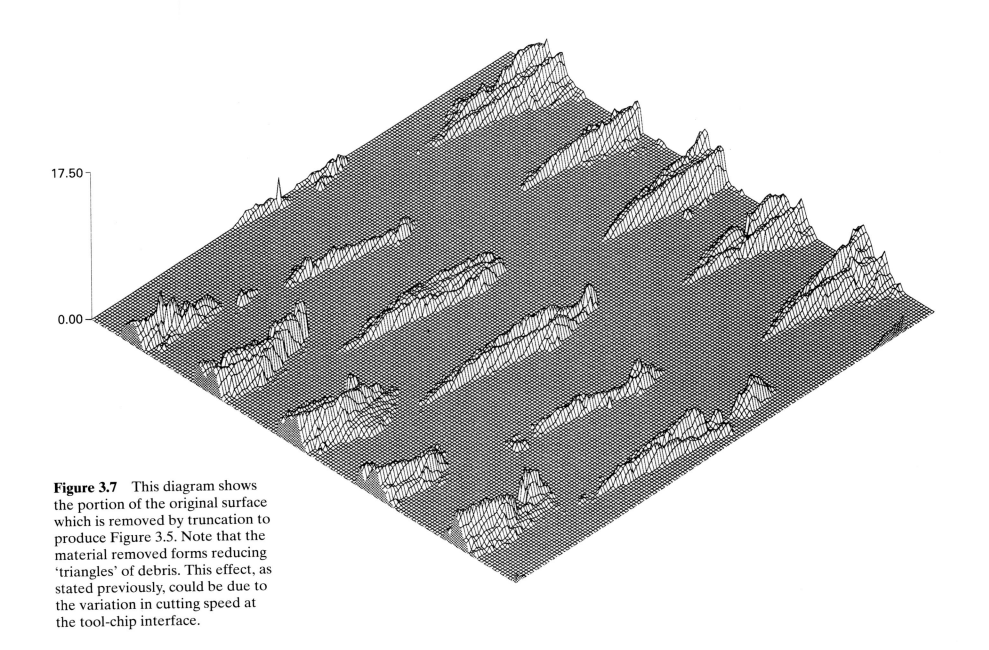

17.50 ─

0.00 ─

Figure 3.7 This diagram shows the portion of the original surface which is removed by truncation to produce Figure 3.5. Note that the material removed forms reducing 'triangles' of debris. This effect, as stated previously, could be due to the variation in cutting speed at the tool-chip interface.

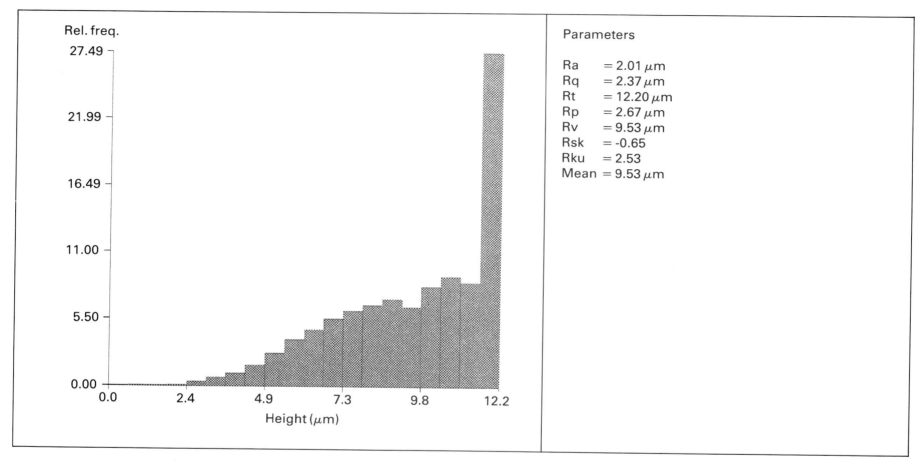

Figure 3.8 The parameters of the surface at 30% *Rt* truncation.
Note that the average roughness parameter, *Ra*, has reduced
considerably and the effect of this can be seen in the height
distribution. The surface has become much more negatively
skewed although the kurtosis value remains unchanged.

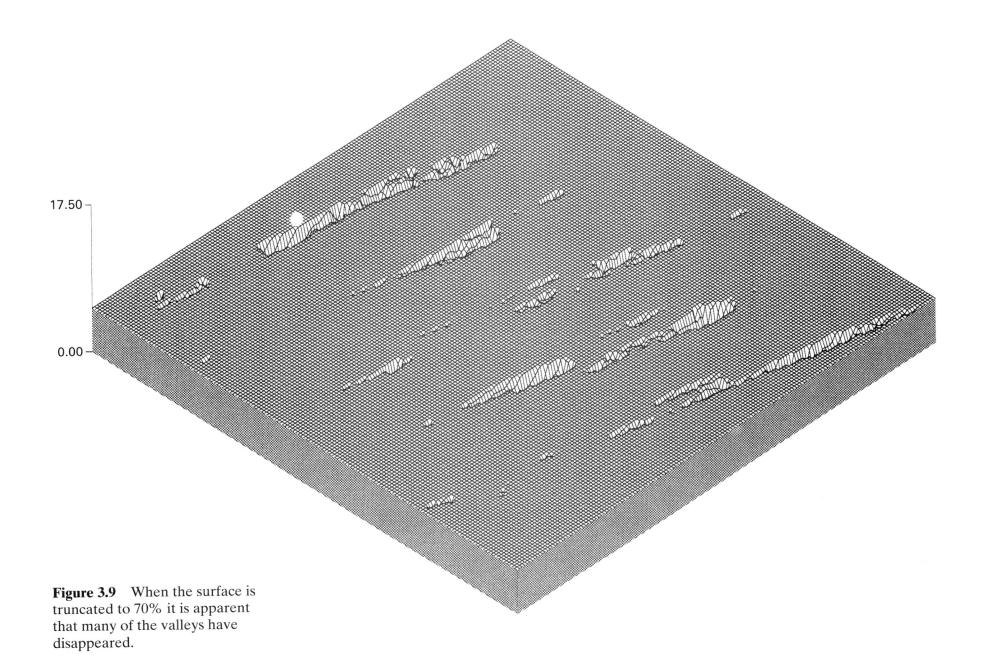

17.50 —

0.00 —

Figure 3.9 When the surface is truncated to 70% it is apparent that many of the valleys have disappeared.

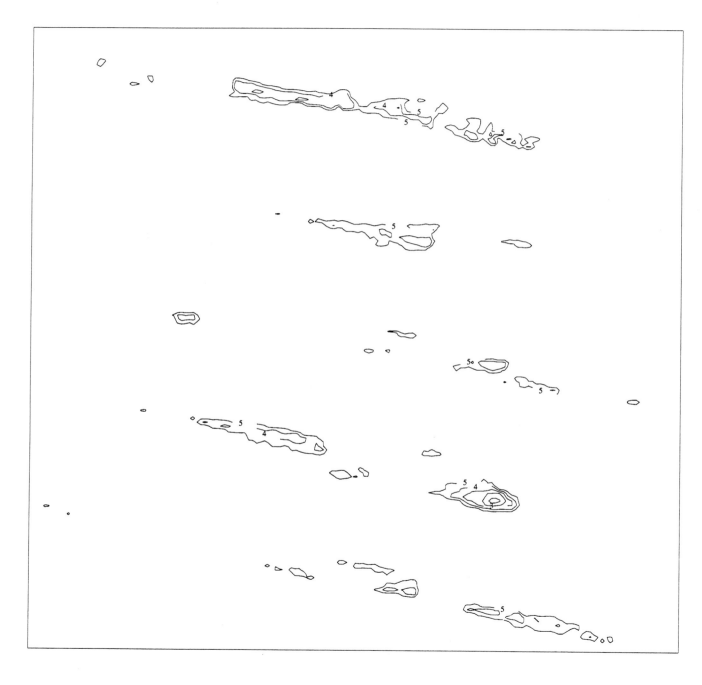

Contour key (μm)

1 : 0.52
2 : 1.56
3 : 2.60
4 : 3.64
5 : 4.68

Figure 3.10 The contour map of the surface after 70% *Rt* truncation shows that the valleys which do remain are evenly distributed. In fact, their separation relates to the feed-rate during machining.

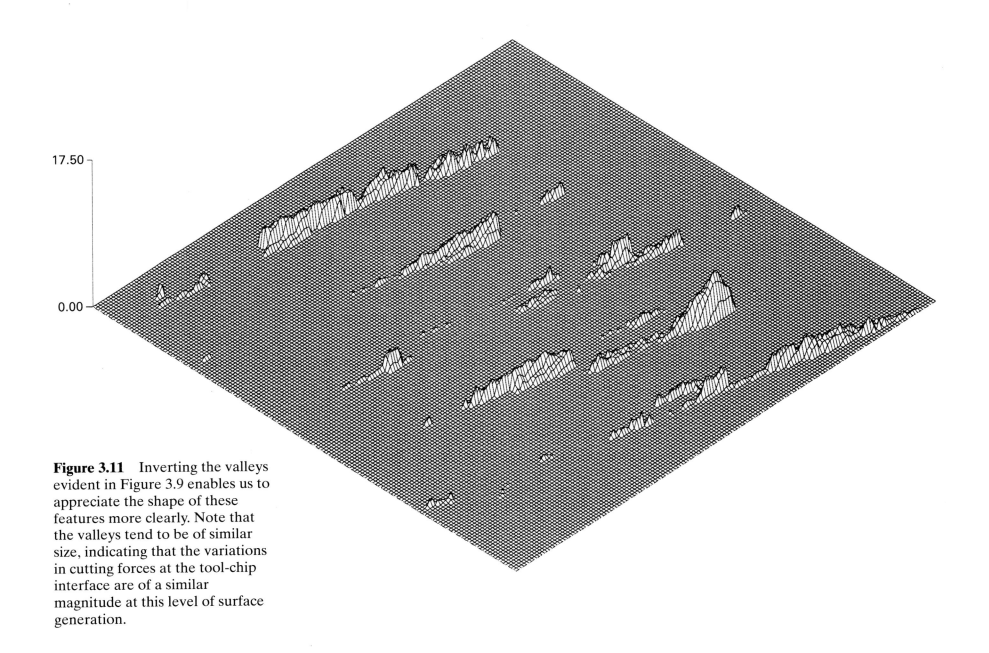

17.50

0.00

Figure 3.11 Inverting the valleys evident in Figure 3.9 enables us to appreciate the shape of these features more clearly. Note that the valleys tend to be of similar size, indicating that the variations in cutting forces at the tool-chip interface are of a similar magnitude at this level of surface generation.

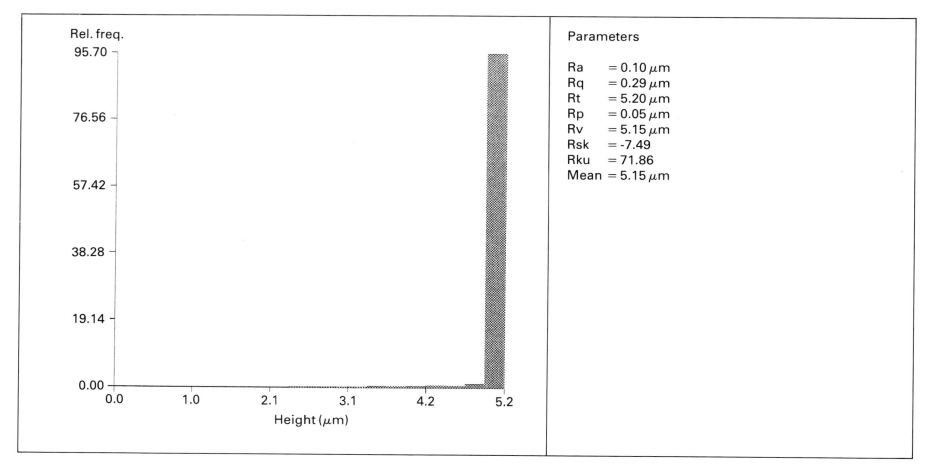

Figure 3.12 The height distribution corresponding to the 70% truncated surface shown in Figure 3.9. As would be expected, most of the surface (95.7%) is now at the uppermost level. This indicates that such a surface, reduced to this level of topography, is unlikely to perform well in service in a tribological environment. This view is supported by the high negative skewness and kurtosis values ($Rsk = -7.49$; $Rku = 71.86$).

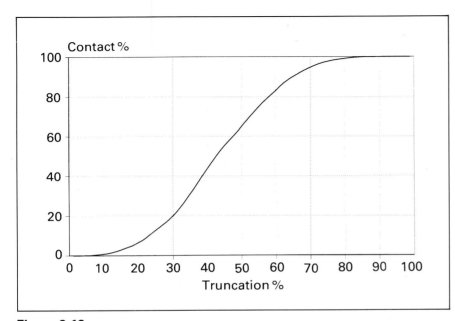

Figure 3.13

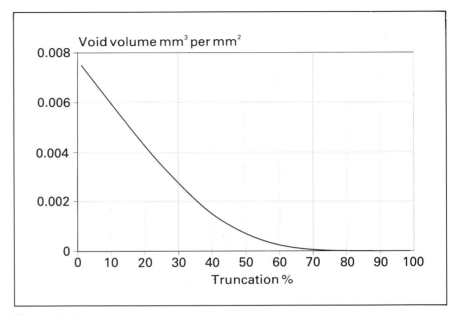

Figure 3.14

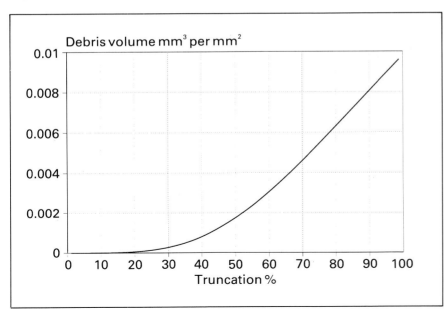

Figure 3.15

Figure 3.13 A plot of contact % against truncation % for the end milled surface shows the nature of the S-shaped curve which indicates that the surface contact area initially increases very rapidly with truncation, reaching 20% at 30% Rt truncation and increases to 80% at 60% Rt truncation.

Figure 3.14 This graph demonstrates the demise of void volume as the surface is truncated. The graph shows that the void volume decreases to 10% of its original value at 47% Rt truncation.

Figure 3.15 A useful method of assessing the functional capabilities of the surface is to examine debris volume versus truncation %. It can be seen that very little debris is generated in the initial stages of truncation (up to approximately 30% Rt). From around the 44% Rt truncation level, however, the debris volume increases in a linear manner through all subsequent truncation levels.

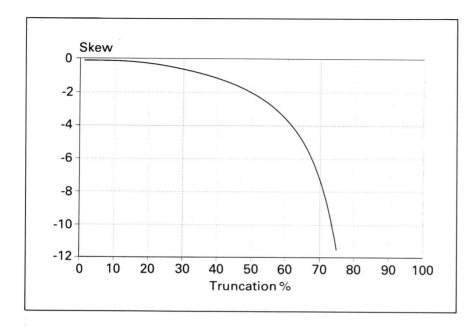

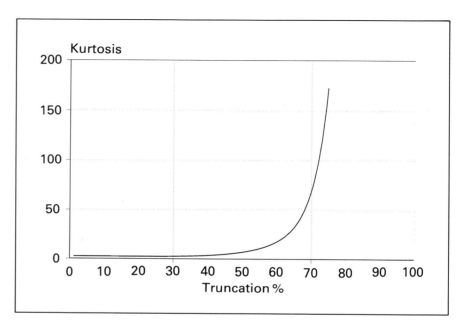

Figure 3.16 This figure represents the progression of the skewness parameter of the 3-D surface. The skewness is marginally negative at first, increasing steadily as truncation increases and reaching a value of *Rsk* = –22 at 80% *Rt* truncation. This value is extremely high and is not typical of other engineering surfaces.

Figure 3.17 This graph is of kurtosis plotted against truncation %. The kurtosis parameter is mathematically related to skewness and high values of skewness thus lead to high values of kurtosis. The reason for such extremes in skewness and kurtosis is related to the shape and regularity of the deepest valleys.

4 GROUND SURFACE

The figures presented in this section show a surface typical of those produced by the fine grinding process. The axonometric projection (Figure 4.1) clearly shows the grinding marks caused by the cutting action of the grits attached to the periphery of the grinding wheel and the directional nature of the grinding process. The grits are held at a variety of levels in relation to the circular envelope scribed by the grinding wheel as it rotates. Some grits are newly formed and sharp, generating a clean cutting action, whilst others exhibit varying degrees of bluntness, leading to surface formation by wear, ductile deformation and fatigue. The size of the grinding ridges and troughs are related to the feed-rate of the dressing diamond as it is fed across the wheel to dress and recondition the cutting edges. The general expectation is that a ground surface is the product of a series of random events. In practice, this is far from being the case, since an orientation exists as a result of the action of the dressing diamond as it prepares the surface of the grinding wheel.

The sample of a ground surface presented in this section is dominated by the appearance of 'gouge' marks, or structured valleys. Such features appear fairly regularly on fine ground surfaces and are not random events. They are caused by sharp grits embedded into the surface of the wheel cutting their form into the surface. This action only occurs in fine grinding, often during spark out, when sharp new grits are available at the uppermost levels of the grinding wheel. As the wheel is rotated at high speed most individual features of the cutting process are removed by a statistical ensemble of actions, which include cutting, wearing, tearing, plastic flow and fracture, all simultaneously occurring over the surface. Hence, in general, the grinding process produces a Gaussian surface, occasionally disrupted by the action of single sharp grits lodged or welded on the surface of the wheel.

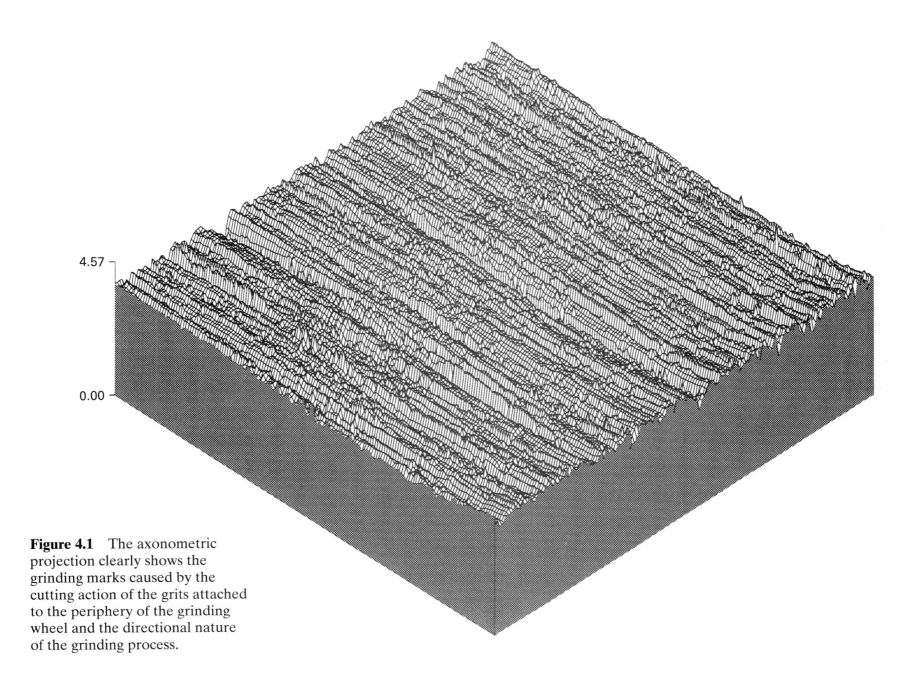

4.57 —

0.00 —

Figure 4.1 The axonometric projection clearly shows the grinding marks caused by the cutting action of the grits attached to the periphery of the grinding wheel and the directional nature of the grinding process.

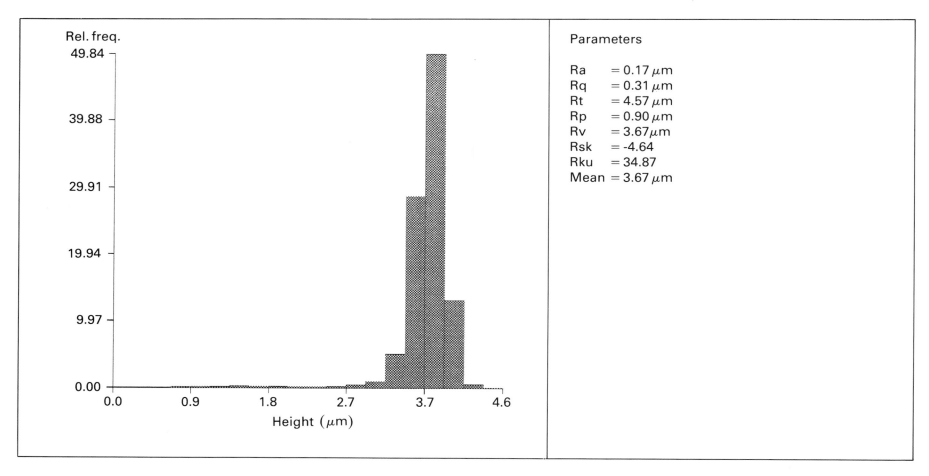

Figure 4.2 The height distribution relating to this surface is typical of that seen from many ground surfaces. Note that the surface is very finely machined ($Ra = 0.17$ μm; $Rt = 4.57$ μm). The histogram also indicates that only a very small portion of the area, probably less than 2%, is concentrated in deep valleys, the distribution of which is not obvious from Figure 4.1.

The distribution of asperities on the surface examined here is highly negatively skewed ($Rsk = -4.64$; $Rku = 34.87$) and in this instance is directly related to the 'gouge' features caused by the action of welded grits.

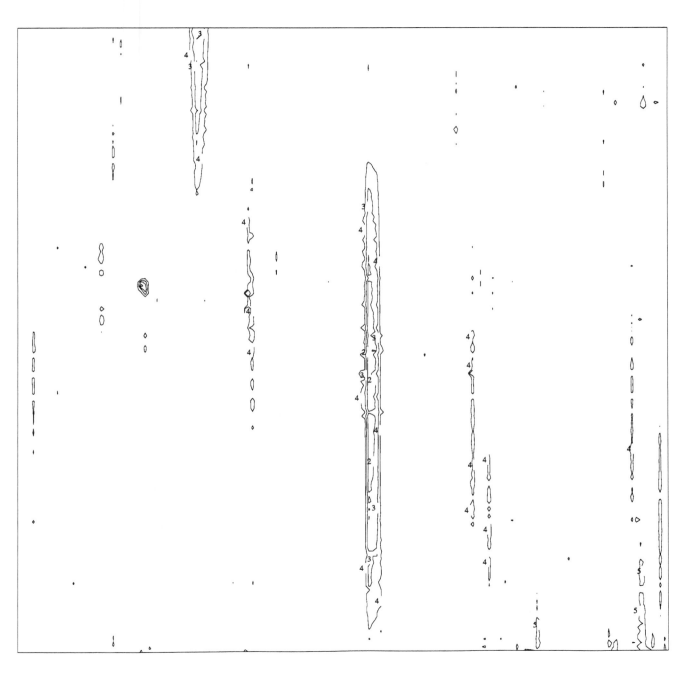

Contour key (μm)

1 : 0.46
2 : 1.37
3 : 2.28
4 : 3.20
5 : 4.11

Figure 4.3a The contour map of the surface starts to reveal the source of these interesting features. Clearly a number of deep valleys exist which are not obvious in Figure 4.1. Note that this contour map does not reveal the micro-roughness well; to achieve this additional contour levels must be plotted and this is demonstrated in the next diagram (Figure 4.3b) on the following page.

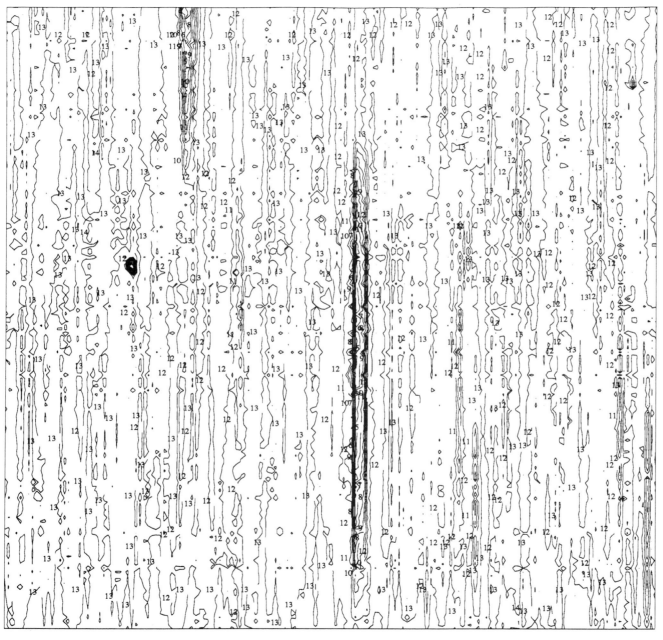

Contour key (μm)

1 : 0.15
2 : 0.46
3 : 0.76
4 : 1.07
5 : 1.37
6 : 1.68
7 : 1.98
8 : 2.29
9 : 2.59
10 : 2.89
11 : 3.20
12 : 3.50
13 : 3.81
14 : 4.11
15 : 4.42

Figure 4.3b With additional contour levels plotted, the micro-roughness of the surface becomes clearly evident.

Figure 4.4 (over) Information on 2-D profiles extracted at orthogonal directions from the surface shown in Figure 4.1. The diagrams indicate that the surface has significantly directional (anisotropic) properties since the distributions relating to the two orthogonal directions of assessment show significant differences. The differences are so large that they cannot be explained by the deep grooves which appear in the surface, and which are discussed in the text.

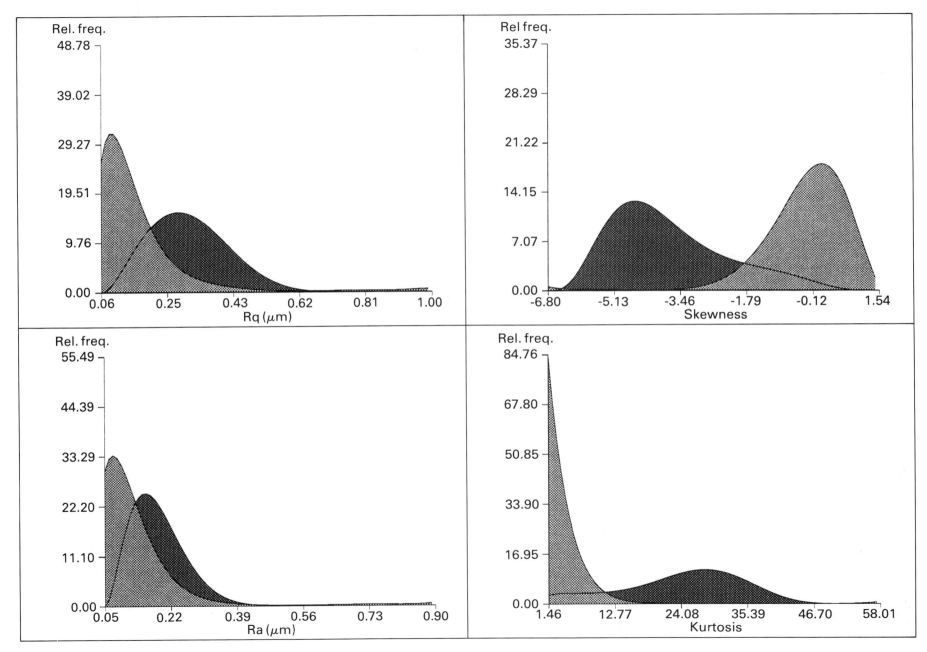

Figure 4.4

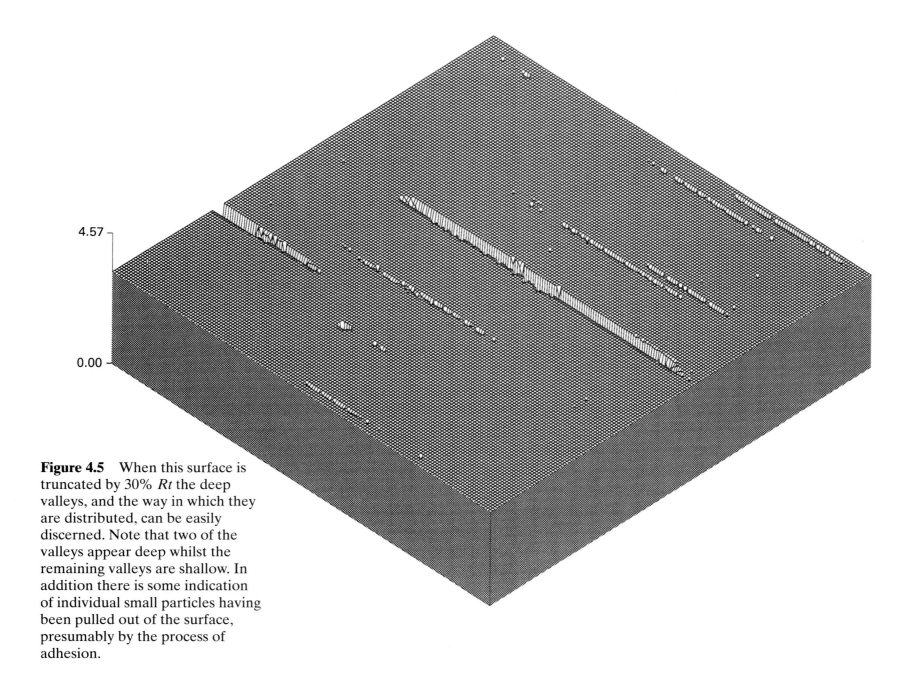

4.57

0.00

Figure 4.5 When this surface is truncated by 30% *Rt* the deep valleys, and the way in which they are distributed, can be easily discerned. Note that two of the valleys appear deep whilst the remaining valleys are shallow. In addition there is some indication of individual small particles having been pulled out of the surface, presumably by the process of adhesion.

Contour key (μm)

1 : 0.32
2 : 0.96
3 : 1.60
4 : 2.24
5 : 2.88

Figure 4.6 The contour map of the truncated surface reinforces the evidence for the views expressed about the surface on the basis of the visualization in Figure 4.5.

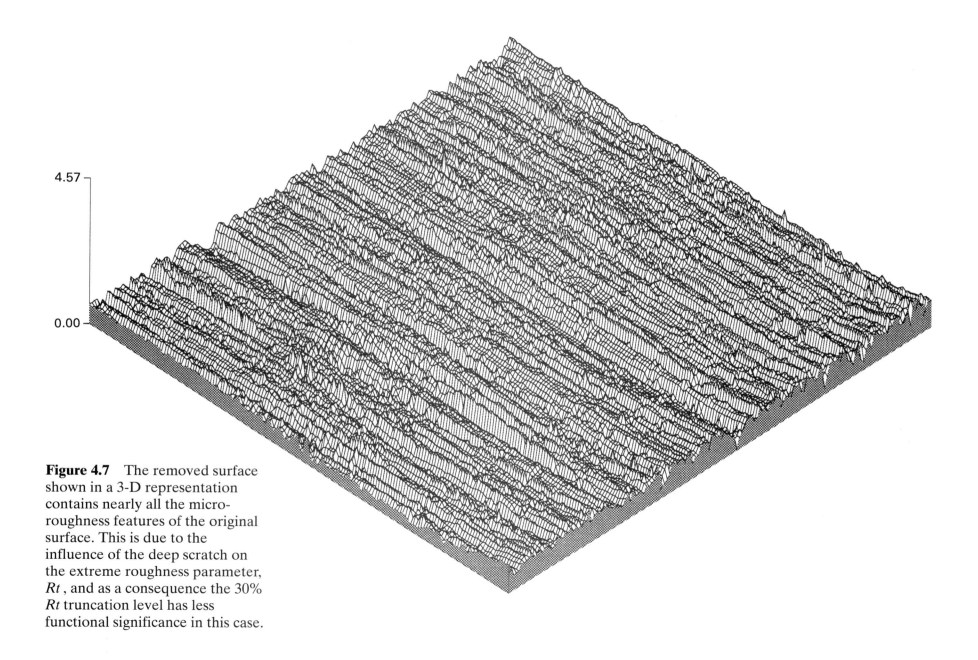

4.57

0.00

Figure 4.7 The removed surface shown in a 3-D representation contains nearly all the micro-roughness features of the original surface. This is due to the influence of the deep scratch on the extreme roughness parameter, Rt, and as a consequence the 30% Rt truncation level has less functional significance in this case.

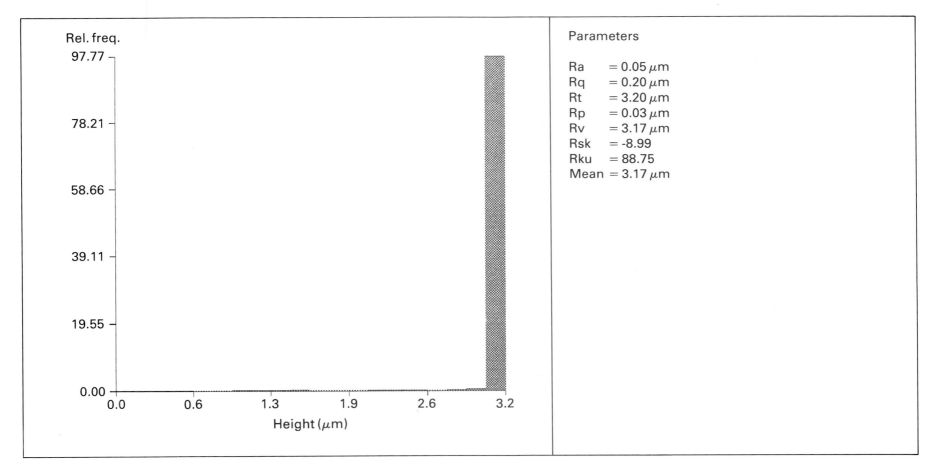

Figure 4.8 The height distribution parameters of the 30% *Rt* truncated surface presented confirms the comments made on the previous figure. Note that skewness is now –8.99 with a kurtosis of 88.75, yielding features that would lead to scuffing if the surface was used in high-load, high-speed applications.

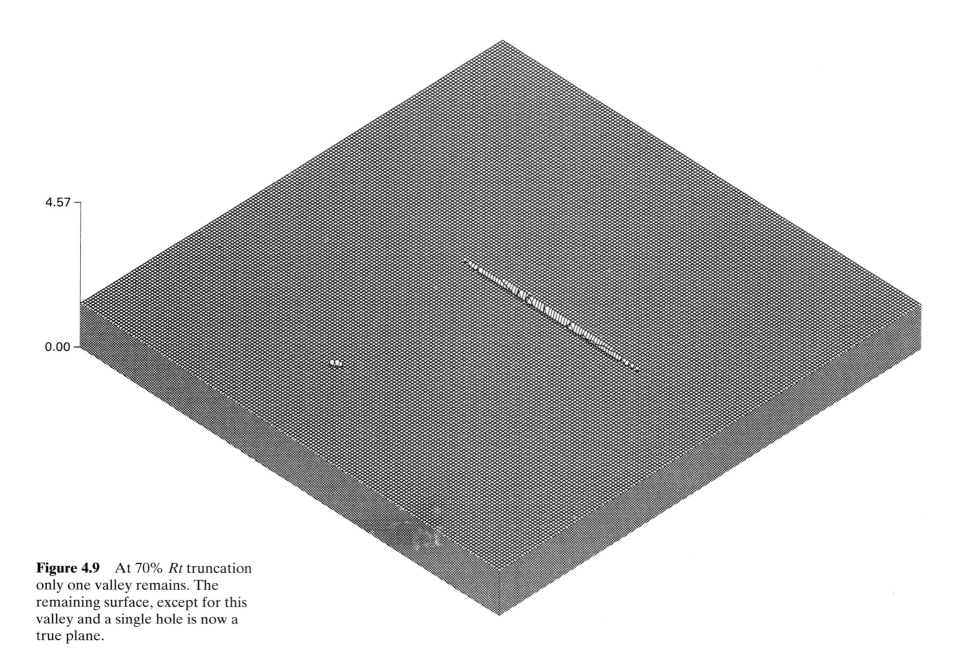

4.57

0.00

Figure 4.9 At 70% *Rt* truncation only one valley remains. The remaining surface, except for this valley and a single hole is now a true plane.

Contour key (μm)

1 : 0.14
2 : 0.41
3 : 0.68
4 : 0.96
5 : 1.23

Figure 4.10   Contour map of the truncated surface as shown in Figure 4.9.

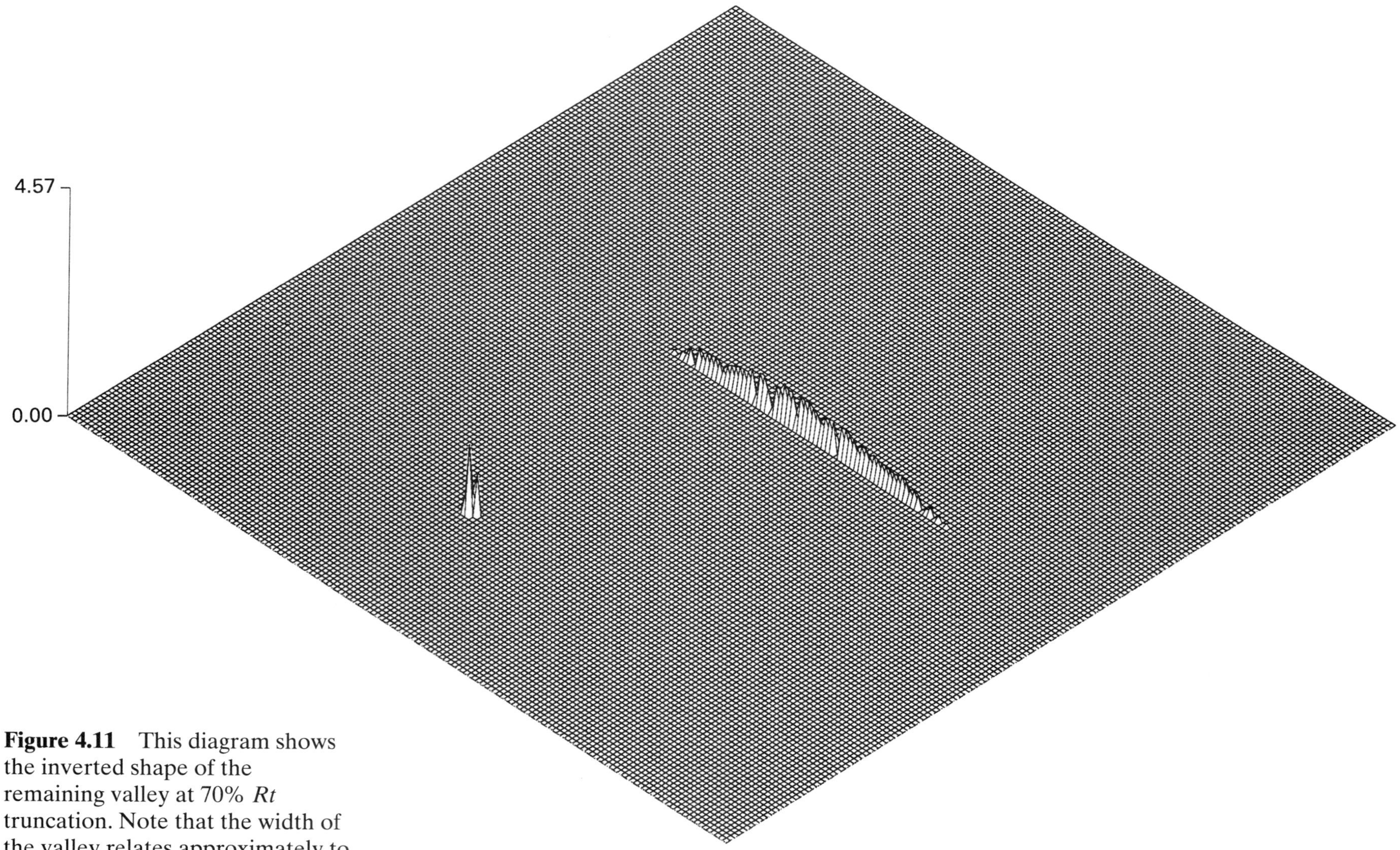

Figure 4.11 This diagram shows the inverted shape of the remaining valley at 70% *Rt* truncation. Note that the width of the valley relates approximately to the width of a typical single grit on a grinding wheel surface.

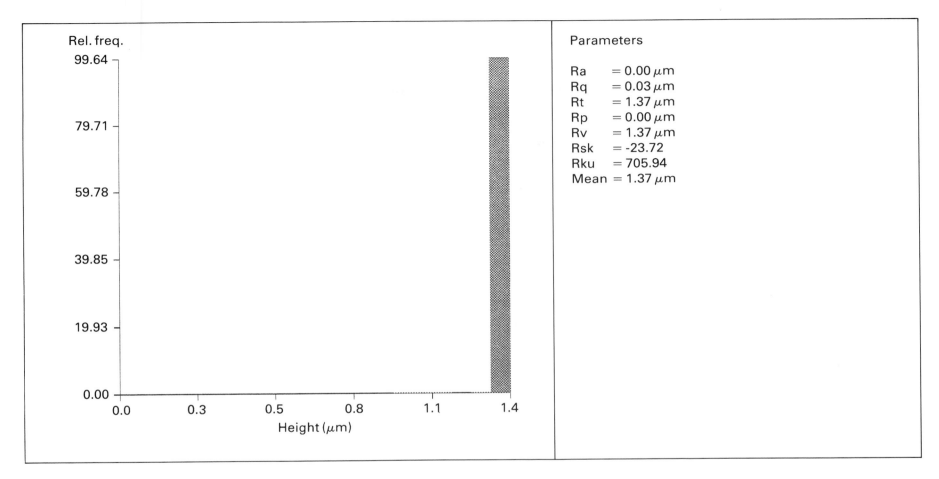

Figure 4.12a The height distribution parameters are becoming unreliable at this level. The total peak to valley roughness $Rt = 1.37$ μm is accurate but note the average roughness value, $Ra = 0.00$ μm (to 2 decimal places).

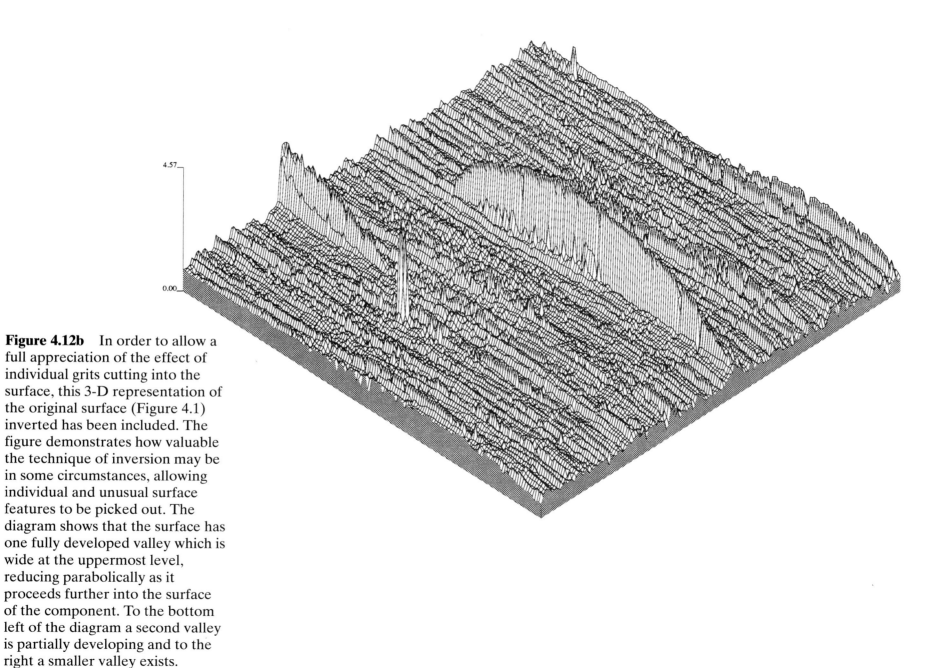

Figure 4.12b In order to allow a full appreciation of the effect of individual grits cutting into the surface, this 3-D representation of the original surface (Figure 4.1) inverted has been included. The figure demonstrates how valuable the technique of inversion may be in some circumstances, allowing individual and unusual surface features to be picked out. The diagram shows that the surface has one fully developed valley which is wide at the uppermost level, reducing parabolically as it proceeds further into the surface of the component. To the bottom left of the diagram a second valley is partially developing and to the right a smaller valley exists.

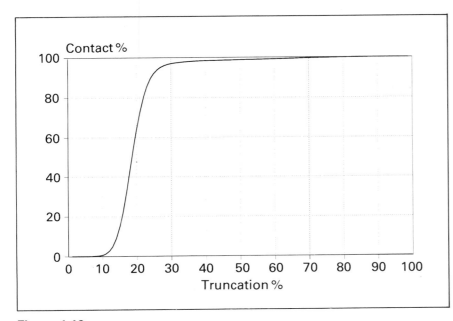

Figure 4.13

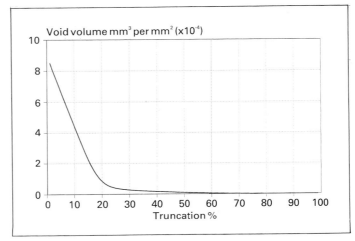

Figure 4.14

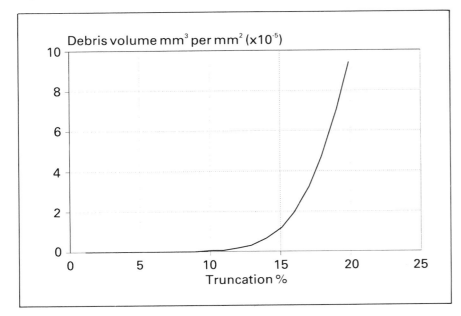

Figure 4.15

Figure 4.13 A plot of contact % against truncation % for the ground surface. The curve shows that there is a rapid transition from low contact to high contact %. The first turning point is at approximately 12%. The second turning point also occurs at a low truncation percentage (27%). The first rapid transition is typical of most ground surfaces but the low value for the second turning point is a result of the presence of the single deep valley which is artificially extending the depth of the surface. Notice that this valley leads to a linear increase in contact % once it becomes the most significant feature at the 30% Rt truncation level.

Figure 4.14 A graph of void volume plotted against truncation level. The reduction is linear until the 18% truncation level. This is an indication of the Gaussian nature of the curves at this level. At approximately 23% the void volume reduces to less than 5% of its original value and becomes almost zero as the one remaining valley is reduced.

Figure 4.15 A graph showing how debris accumulates as the surface is truncated. After the 20% Rt truncation level the rate of material removal is approximately constant, as only the occasional scratch remains. As a result, for clarity, only the initial 20% Rt truncation is presented.

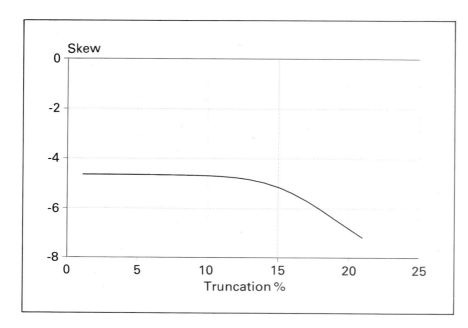

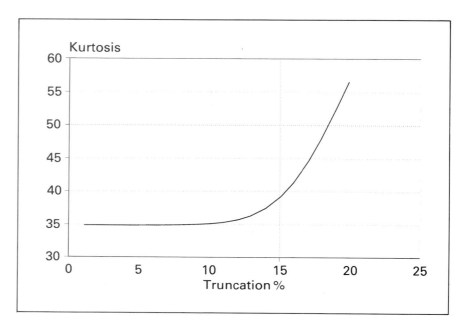

Figure 4.16 The progression of skewness plotted against truncation % up to a 25% *Rt* truncation level. It can be seen that skewness remains approximately constant (*Rku* = –4.7) until a 10% *Rt* truncation level is reached. Its value then increases gradually until 22% *Rt* truncation and rises dramatically after this point.

Figure 4.17 The behaviour of the kurtosis parameter up to the 20% *Rt* truncation level. It can be seen that the initial value of *Rku* = 35 remains constant until the 10% *Rt* truncation level and then rises gradually until a constant rate is obtained from approximately the 17% *Rt* truncation level onwards.

5 SAND BLAST SURFACES

The figures presented in this section relate to a surface typical of those produced by the sand blasting process. The surface is generated by particles of approximately similar size impinging on the surface at random positions and orientations. The impact leads to high kinetic energy being imparted to the surface, causing localized heating and, in many cases, asperity melting. The force of impact also blasts particles of the surface away from the main body of the material. It is probable that the kinetic energy of each individual event is approximately constant. Thus a number of random events clustering in similar positions on the surface – a statistically likely occurrence – will lead to the formation of larger valleys of the kind that become evident in this surface when it is subjected to simulated wear. The result is a structure that is random, not only spatially but also in amplitude. Depending upon the size of the blasted particles and their velocity, the surface may be extremely rough, or, alternatively, less rough with a fine overall structure. Irrespective of the amplitude of the roughness encountered, the random nature of the surface would be similar throughout all types of surface generation conditions.

It should be recognized that the roughness, and thus the volume of the voids, in this sample could have been increased if greater intensity sand blasting had been performed, which might have been desirable to increase the wear potential of the surface and to increase the general roughness. In addition it should be recognized that the surface generation method would induce negative residual stresses into the surface, a further useful tribological feature.

The fact that when the surface is subjected to simulated wear there is a gradual transition from low contact to high contact (see Figure 5.13) is a very desirable feature in a tribological application. This is an interesting feature, since the surface is generated by a system of random events typical of those occurring naturally, for example rain drops falling on a pond. It may be that the investigation of surfaces generated by random events will lead to the best surfaces for tribological application, and this view may lead to such generation methods being more extensively investigated.

In general, this surface has some useful properties in a tribological sense and the data would suggest further investigation to determine how effective sand blasting would be as a final finishing technique for a range of tribological interactions.

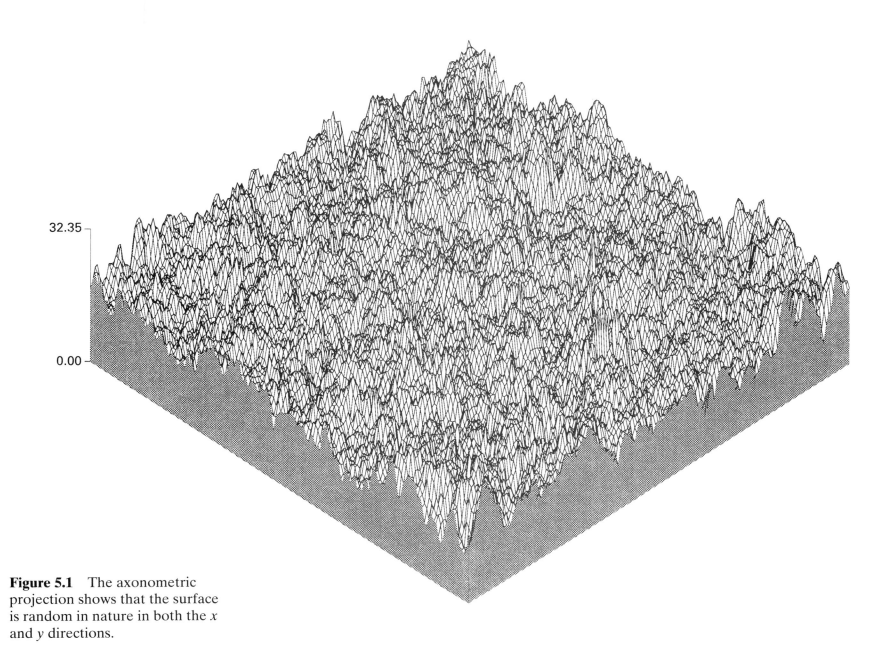

32.35

0.00

Figure 5.1 The axonometric projection shows that the surface is random in nature in both the *x* and *y* directions.

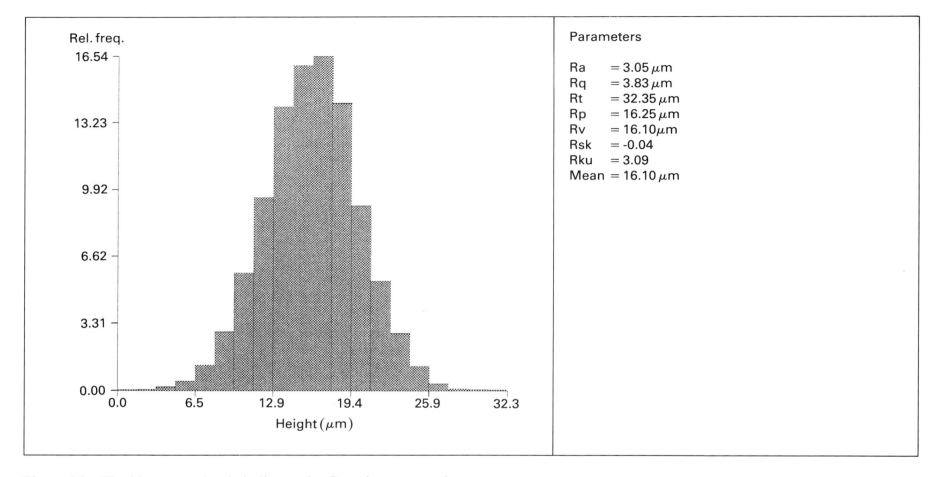

Figure 5.2 The histogram clearly indicates the Gaussian nature of the surface. Note that the maximum relative frequency is centred closely to the mid-point of the distribution of asperity heights. The skewness and kurtosis values ($Rsk = -0.4$; $Rku = 3.09$) are typical of a Gaussian distribution.

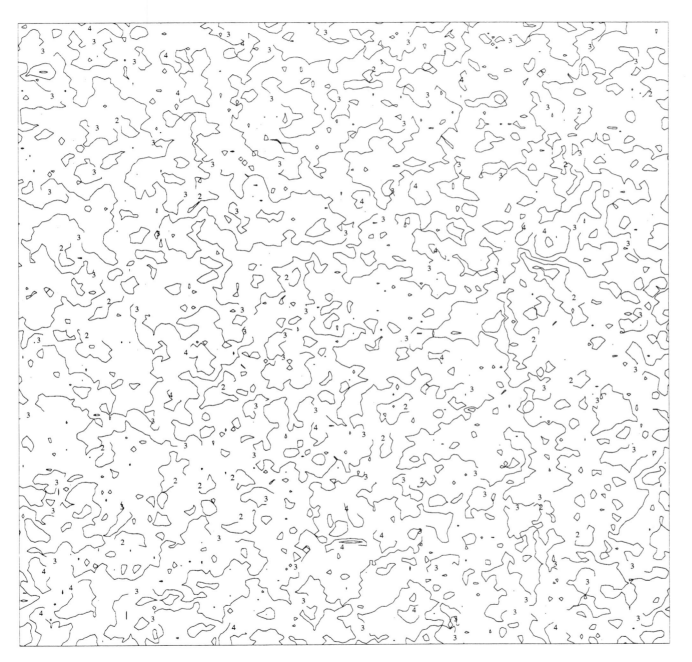

Contour key (μm)

1 : 3.23
2 : 9.70
3 : 16.17
4 : 22.64
5 : 29.11

Figure 5.3 The contour map indicates the random nature of the asperity heights and confirms the visualization presented in Figure 5.1.

Figure 5.4 (over) These four charts present information on 2-D profiles extracted at orthogonal directions from the 3-D surface shown in Figure 5.1. Sand blasted surfaces are claimed to be totally random. A truly random surface would yield distributions which were totally in phase with each other and overlapping completely and the four figures for Ra', Rq' skewness and kurtosis indeed indicate closely similar distribution for all cases. The differences which do exist are primarily due to sampling variations which occur when assessing small samples from a very large population of data, such as the surface under consideration.

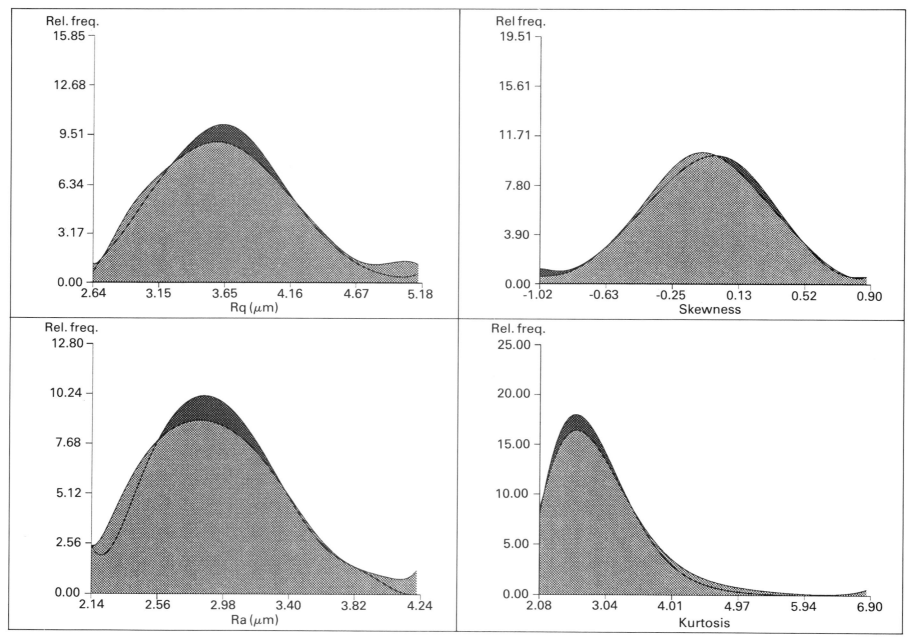

Figure 5.4

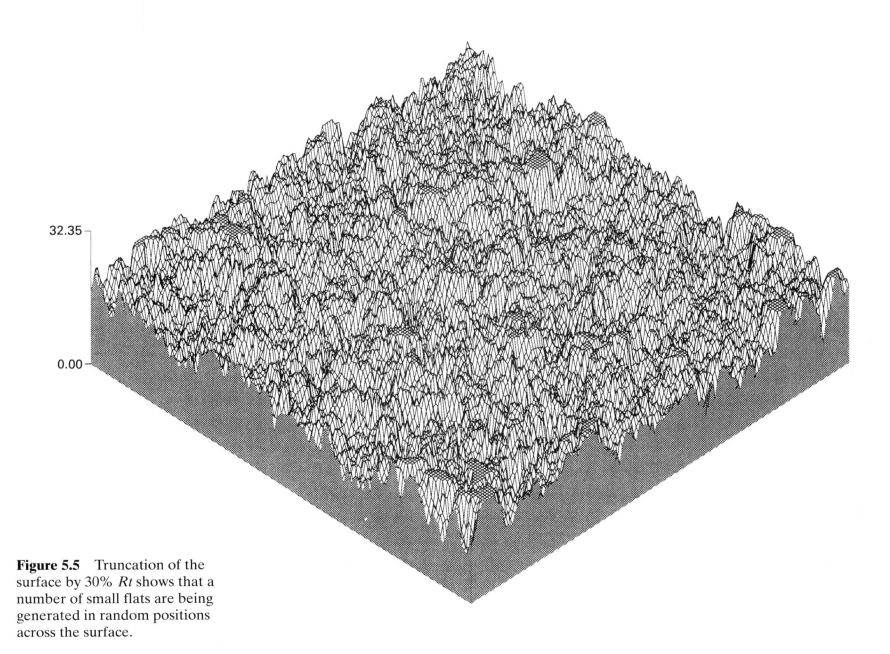

32.35

0.00

Figure 5.5 Truncation of the surface by 30% *Rt* shows that a number of small flats are being generated in random positions across the surface.

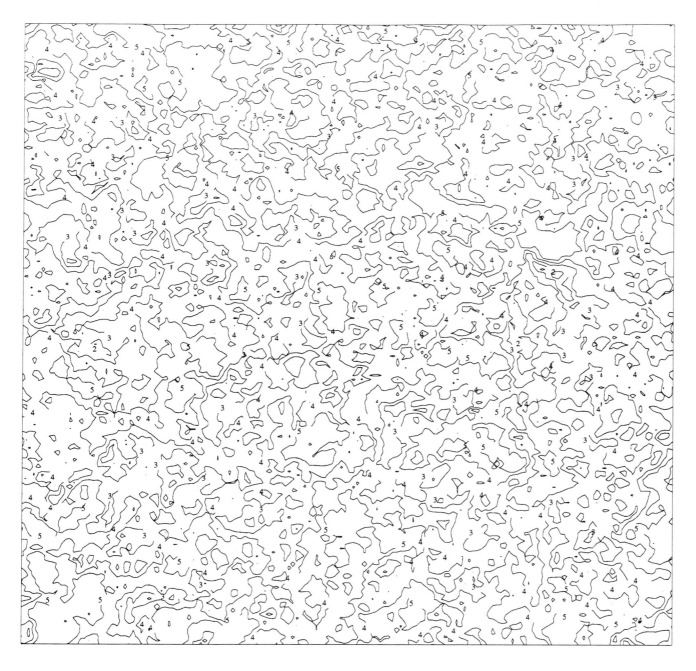

Contour key (μm)

1 : 2.26
2 : 6.79
3 : 11.32
4 : 15.85
5 : 20.38

Figure 5.6 The contour map of the truncated surface makes it clear that the sizes of the flats vary significantly. This is to be anticipated because of the nature of the surface generation technique, where valleys are formed within valleys as the kinetic energy of the impacted particles is imparted.

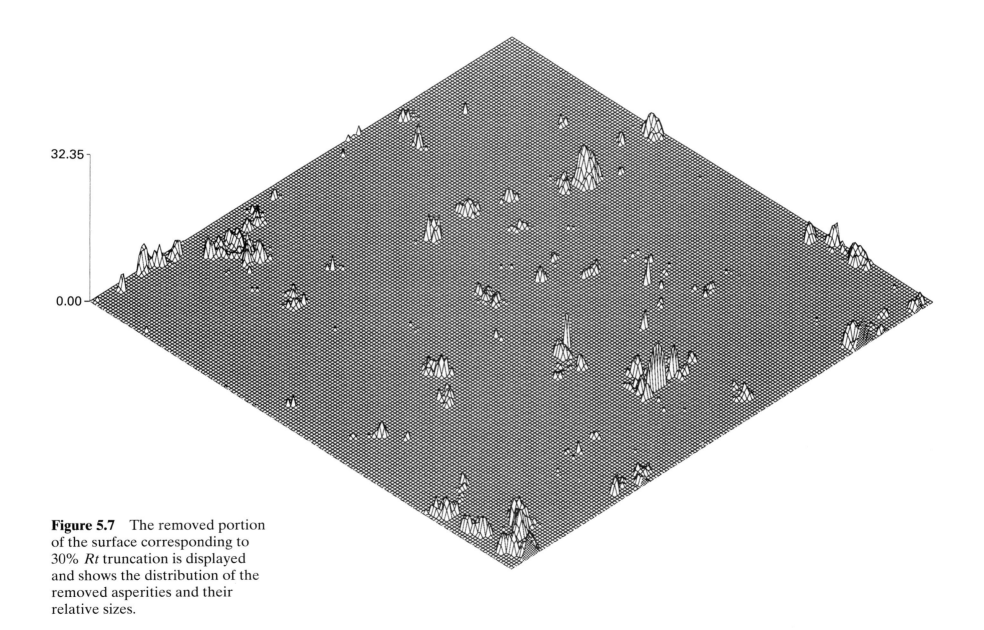

32.35

0.00

Figure 5.7 The removed portion of the surface corresponding to 30% *Rt* truncation is displayed and shows the distribution of the removed asperities and their relative sizes.

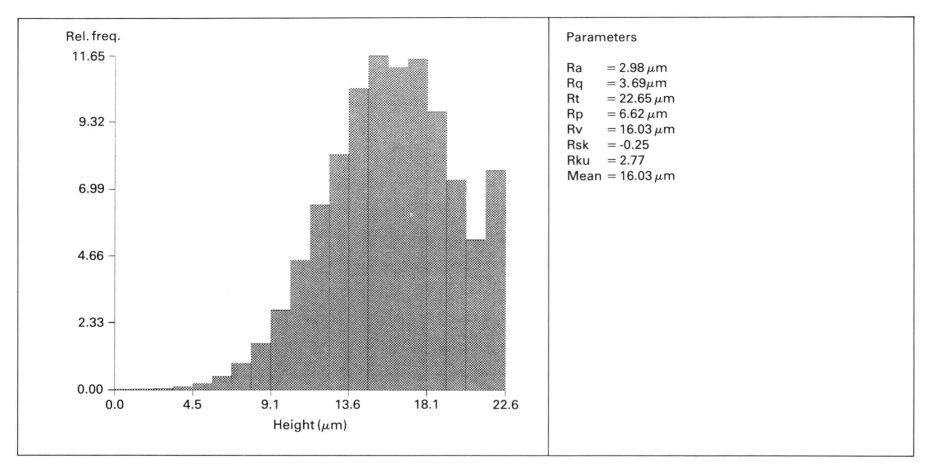

Figure 5.8 Although the height distribution for the truncated surface looks significantly different, the associated height parameters are very similar to those of the original surface (Figure 5.2). The removed part of the distribution has been replaced by an impulse.

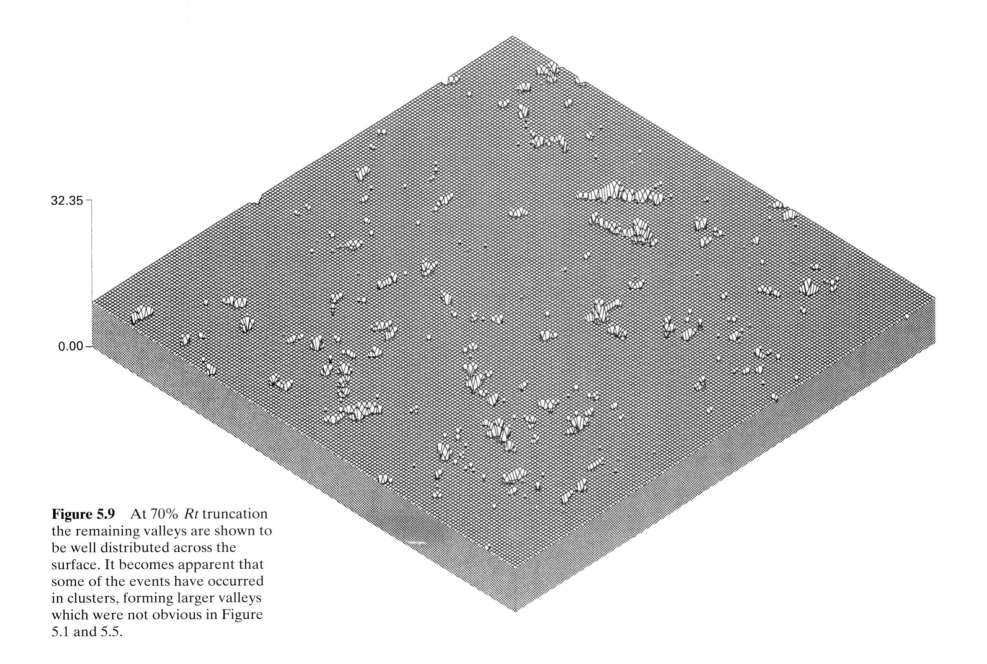

32.35

0.00

Figure 5.9 At 70% *Rt* truncation the remaining valleys are shown to be well distributed across the surface. It becomes apparent that some of the events have occurred in clusters, forming larger valleys which were not obvious in Figure 5.1 and 5.5.

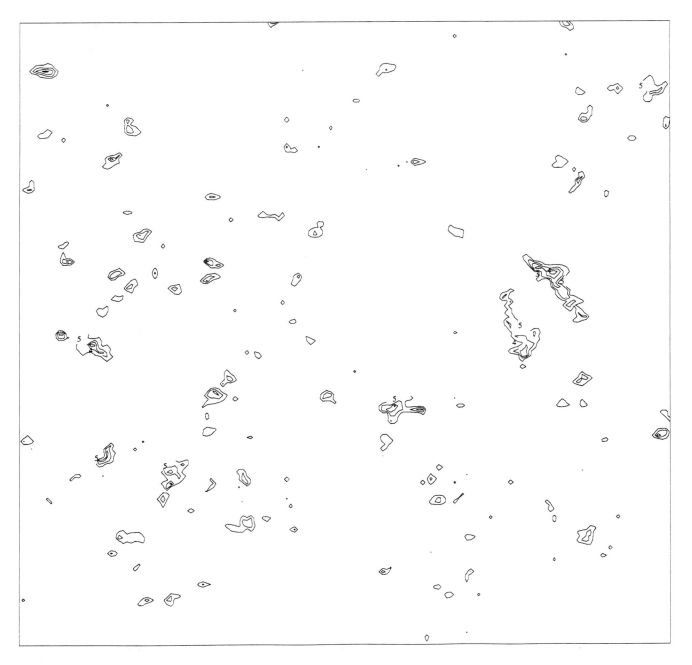

Contour key (μm)

1 : 0.97
2 : 2.91
3 : 4.85
4 : 6.79
5 : 8.73

Figure 5.10 The contour map confirms the even distribution of the valleys seen in Figure 5.9.

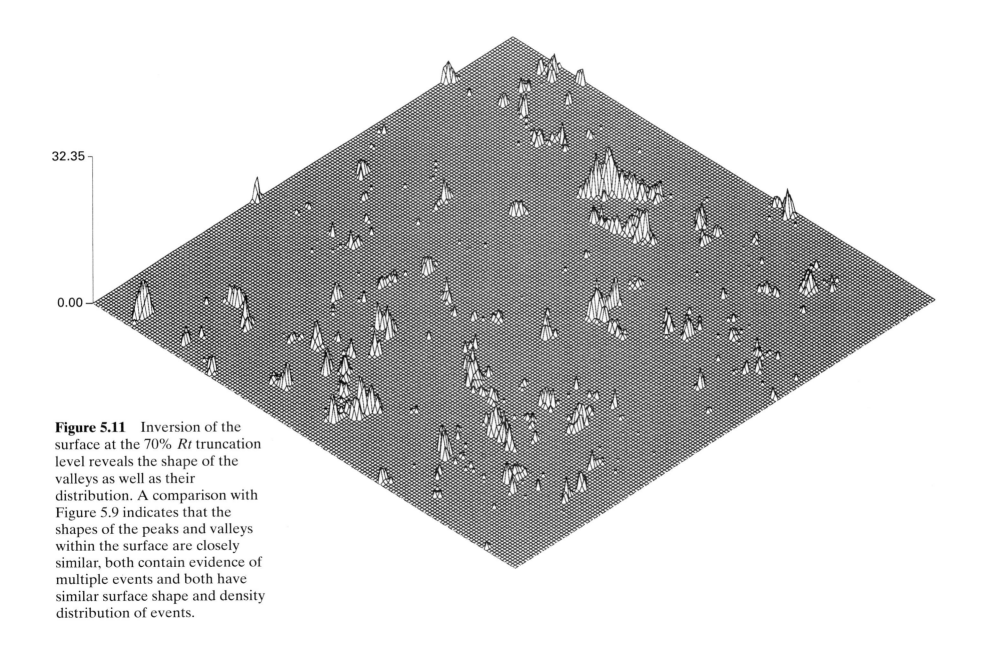

32.35 —

0.00 —

Figure 5.11 Inversion of the surface at the 70% *Rt* truncation level reveals the shape of the valleys as well as their distribution. A comparison with Figure 5.9 indicates that the shapes of the peaks and valleys within the surface are closely similar, both contain evidence of multiple events and both have similar surface shape and density distribution of events.

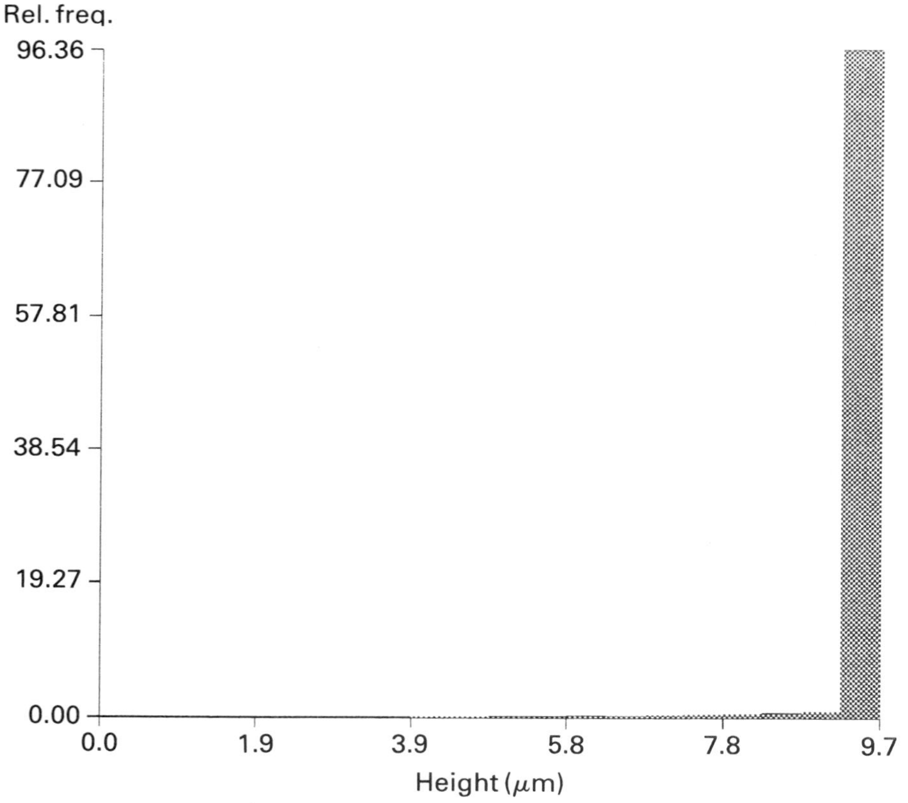

Rel. freq.

Parameters
Ra = 0.15 μm
Rq = 0.49 μm
Rt = 9.70 μm
Rp = 0.08 μm
Rv = 9.62 μm
Rsk = -8.38
Rku = 89.30
Mean = 9.62 μm

Height (μm)

Figure 5.12 At the 70% *Rt* truncation level the distribution of surface heights is overwhelmed by the impulse representing the truncated portion of surface. Both skewness and kurtosis values are extremely high at this level of investigation.

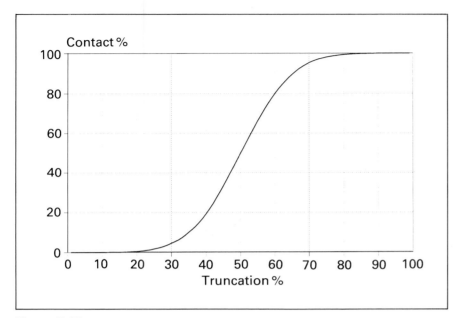

Figure 5.13

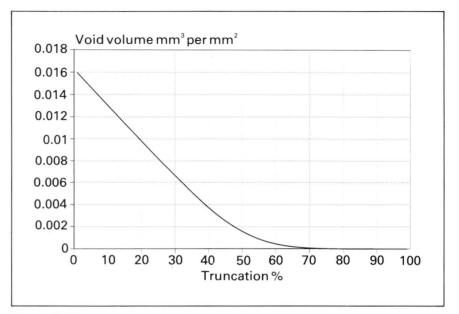

Figure 5.14

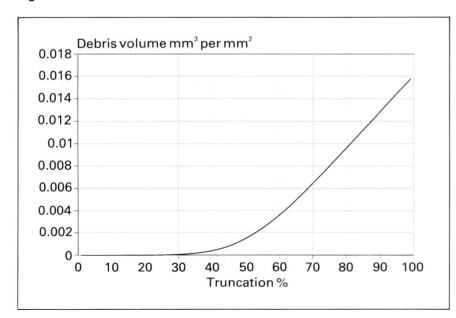

Figure 5.15

Figure 5.13   This graph presents contact % against truncation % for the sand blasted surface. The nature of the curve is different from those shown previously. The curve moves through a steady transition from low contact % to high contact %. It is shown in the diagram that 10% contact occurs at approximately 35% truncation, and that contact rises to 90% at 65% truncation.

Figure 5.14   The graph of void volume plotted against truncation % shows that the void volume reduces linearly from 0% truncation to near 50% Rt truncation, an ideal tribological feature.

Figure 5.15   The reverse of the previous plot, debris volume against truncation % shows that the debris volume is virtually zero until 35% Rt truncation, after which it accelerates, having a near linear increase with truncation at the 50% Rt truncation level.

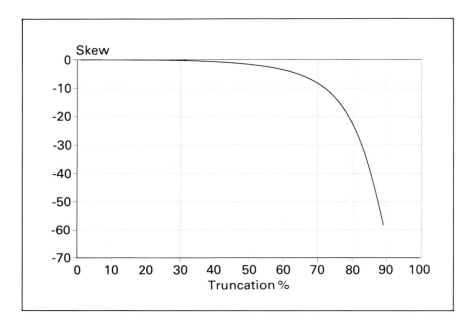

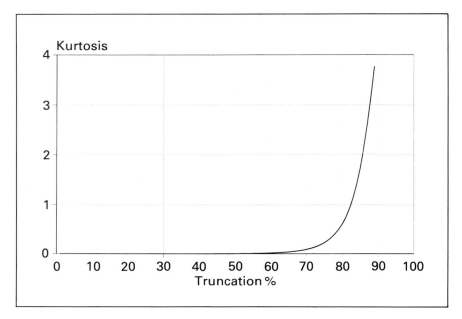

Figure 5.16   The graph shows that the skewness parameter follows its natural trend, increasing towards negative infinity.

Figure 5.17   Like skewness, kurtosis follows its natural trend, which is towards positive infinity. Surfaces will operate in a tribological environment to approximately –8.00 skewness and 20.0 kurtosis values. These values occur at approximately 70% *Rt* truncation.

6 FLY CUT SURFACES

The figures presented in this section relate to a typical surface produced by the fly cut process. The surface has a highly correlated structure, generated by the action of the single-point tool as it revolves over the surface. The correlation is caused by the feed-rate of the machine table as the workpiece is traversed under the rotating cutting tool. Due to the small area under assessment, it is not possible to easily detect the curvature relating to the scribed circumference of the cutting tool as it swings in an arc across the surface. Notice the considerable micro-roughness on Figure 6.1 which is typical of the cutting action associated with single-point cutting tools.

Some of the individual small features are probably caused by tool chatter and small amounts of asperity welding, which is a common feature of single-point machining processes.

When the surface is subjected to simulated wear it can be seen (Figure 6.7) that two bands form in each group of asperities. This may be due to chatter at the tool point. Also, the height of the asperities tends to alter in a cyclic pattern, being large at first and then reducing before becoming large again. This effect may be explained by either the dynamic stiffness of the fly cutting tool or variations in the hardness of the material being cut.

The surface will probably retain good general lubricating properties until approximately 70% of the Rt value has been removed by wear, beyond this level the distribution of surface heights indicates that the tribological properties of this surface are significantly reduced leading to a situation where failure by scuffing is liable to occur.

The combination of parameters (Figure 6.2) is typical of those found from surfaces produced by single-point machining. Similar values have been seen in many other processes such as turning and shaping. This phenomenon is due to the geometry of the cutting tool. The tool shape is designed to give long tool life and to minimize the tool-chip interface temperature. The magnitude of the average roughness parameter, Ra, for fly cutting is less than that found for similar single-point cutting processes, for example turning. The scale of the roughness results from a smaller depth of cut and lower feed-rates used in the fly cutting process than other single point cutting processes. This small depth of cut and lower feed-rate is purely a consequence of the lack of rigidity of the cutting tool in its holder. Rigid tool holding is always difficult to achieve with the fly cutting process.

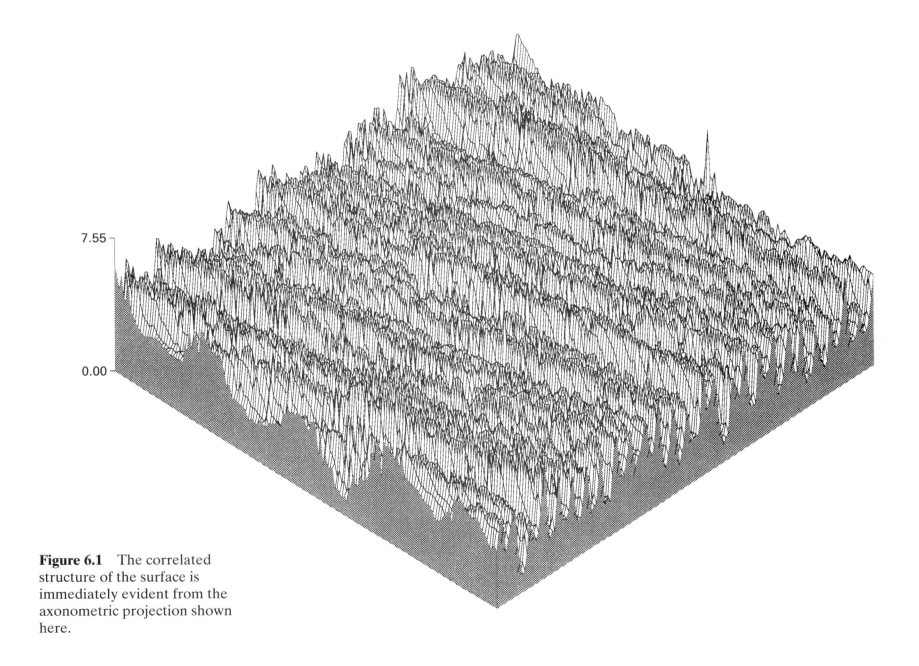

7.55 ⌐

0.00 ⌐

Figure 6.1 The correlated structure of the surface is immediately evident from the axonometric projection shown here.

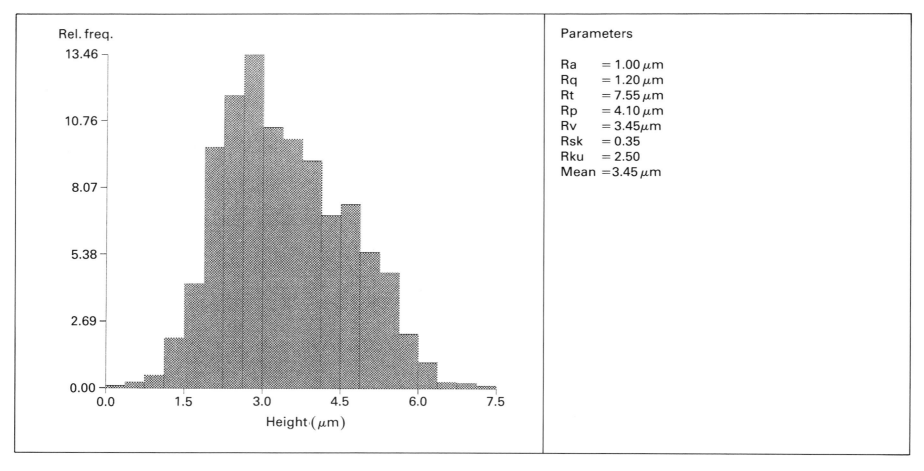

Figure 6.2 The histogram shows that the surface is slightly positively skewed and has a low kurtosis value.

Contour key (μm)

1 : 0.75
2 : 2.26
3 : 3.77
4 : 5.28
5 : 6.79

Figure 6.3 The contour map shows that the cutting action appears at first sight to form a series of parallel grooves, but closer examination will reveal that the grooves are very slightly curved; the magnification used to plot the diagram disguises much of the curvature which is present.

Figure 6.4 (over) The four curves presented here give information obtained by taking individual traces in two orthogonal directions across the 3-D surface under assessment. They provide information on the relative frequency of particular parameters: Rq', Rq', skewness and kurtosis. The diagrams clearly show that the 2-D topography is significantly different when taken in the two directions. The general level of roughness along the lay (which is characterized Ra' and Rq') has a significantly lower value than that across the lay.

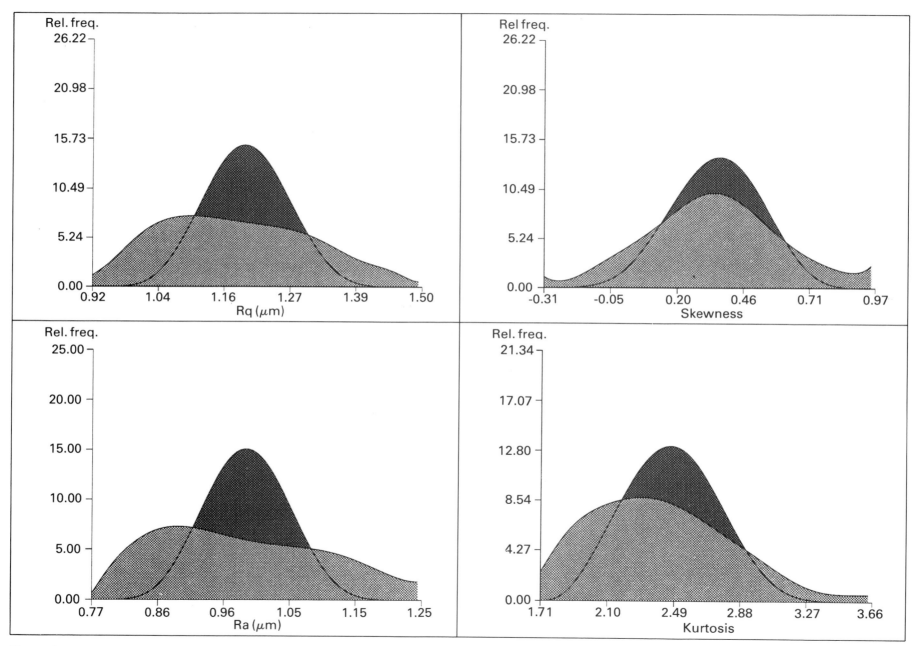

Figure 6.4

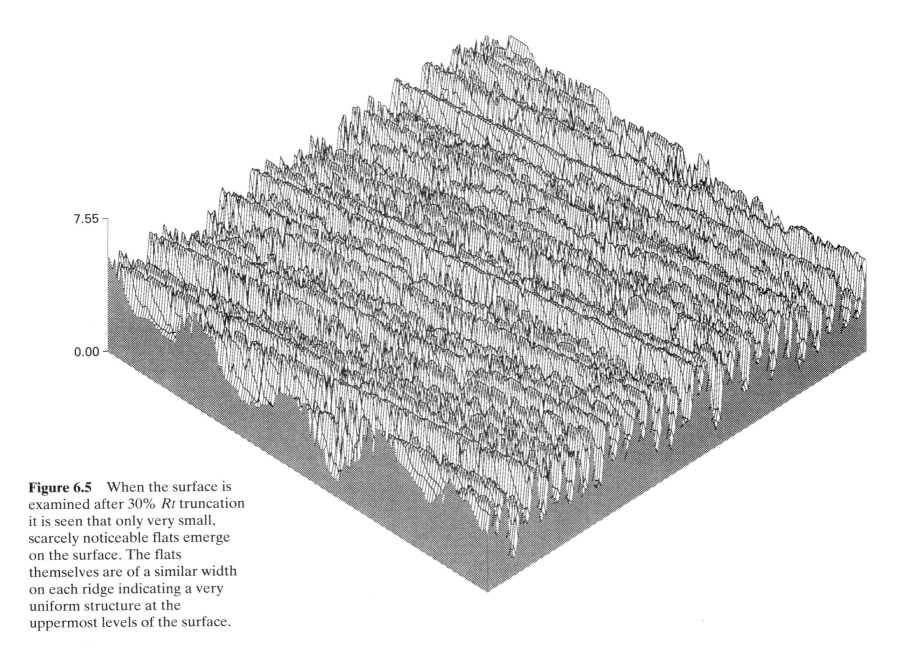

7.55

0.00

Figure 6.5 When the surface is examined after 30% *Rt* truncation it is seen that only very small, scarcely noticeable flats emerge on the surface. The flats themselves are of a similar width on each ridge indicating a very uniform structure at the uppermost levels of the surface.

Contour key (μm)

1 : 0.53
2 : 1.59
3 : 2.65
4 : 3.71
5 : 4.77

Figure 6.6 The contour map of the surface is shown in Figure 6.3. As would be expected, this map differs little from that shown in Figure 6.3. This indicates that very little structural change is occurring at this level of truncation.

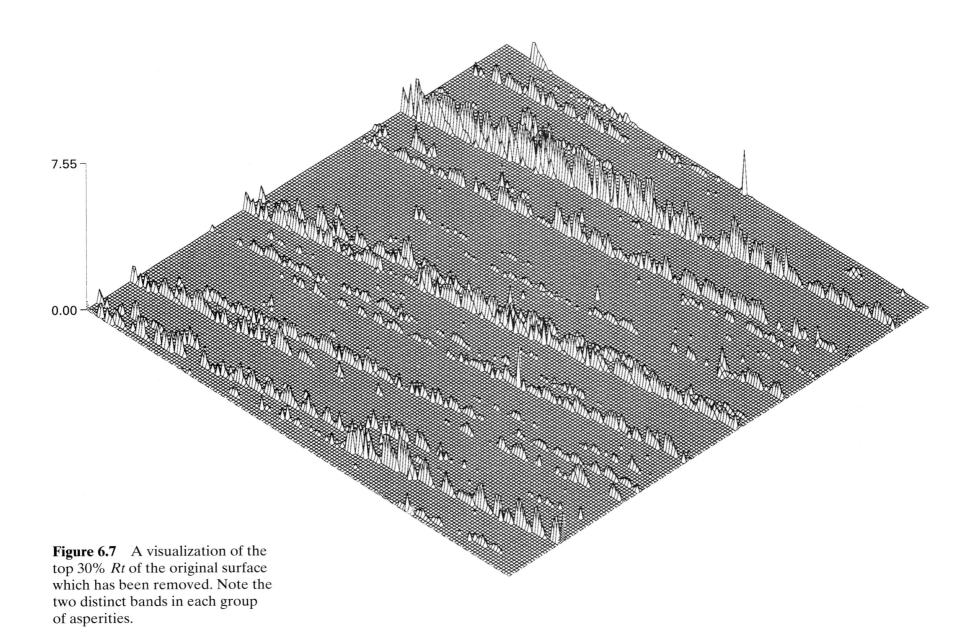

7.55

0.00

Figure 6.7 A visualization of the top 30% *Rt* of the original surface which has been removed. Note the two distinct bands in each group of asperities.

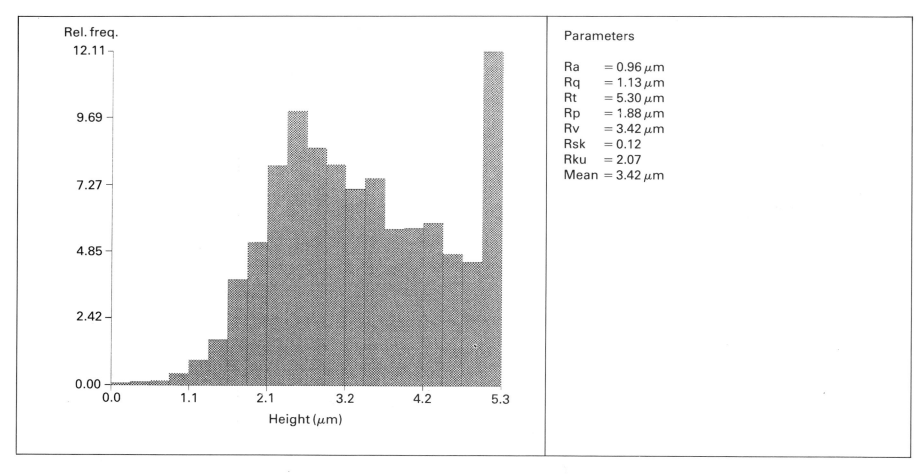

Figure 6.8 The height distribution of the 30% *Rt* truncated surface shows that the skewness value has reduced to *Rsk* = +0.12 and the kurtosis value has also decreased. This again is typical of a surface which has been cut by a single-point cutting tool.

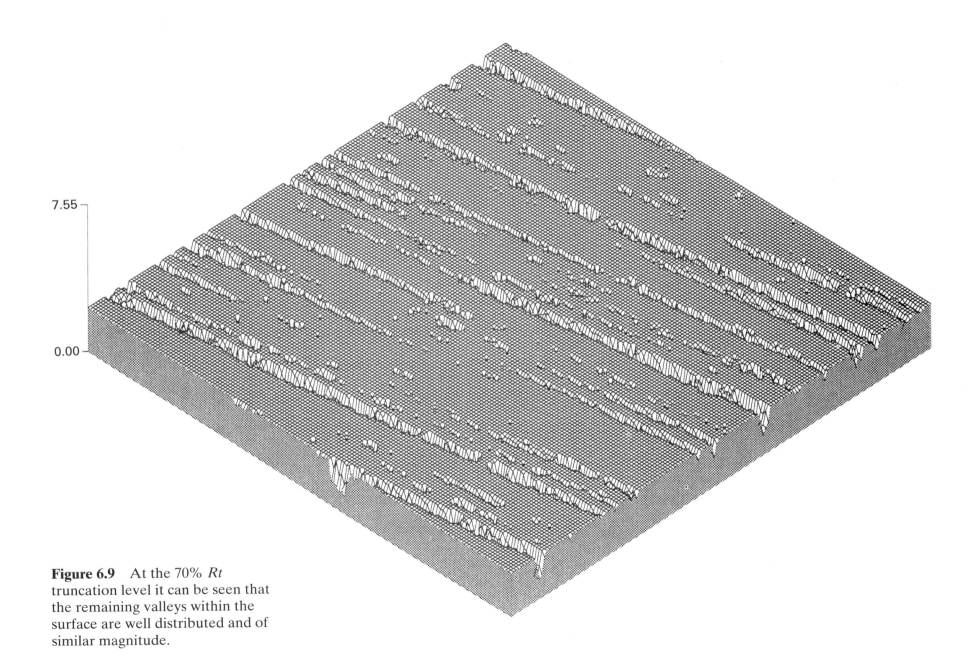

7.55

0.00

Figure 6.9 At the 70% *Rt* truncation level it can be seen that the remaining valleys within the surface are well distributed and of similar magnitude.

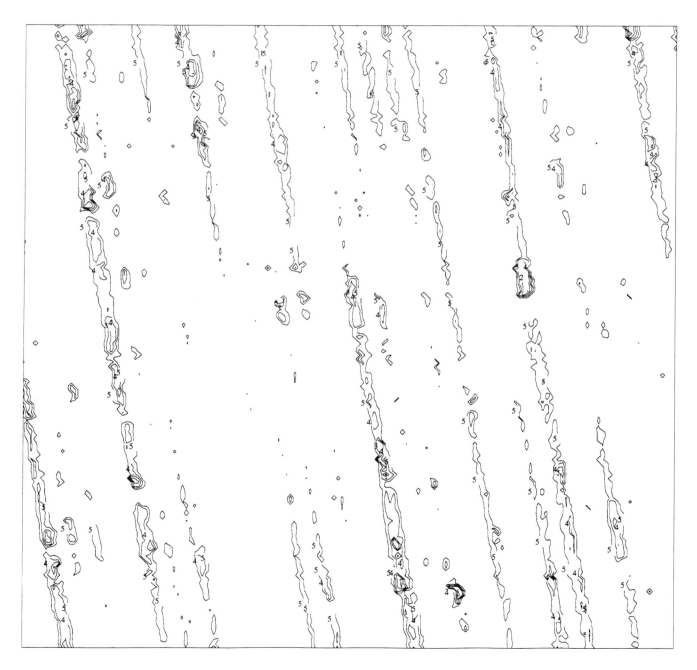

Contour key (μm)

1 : 0.22
2 : 0.67
3 : 1.13
4 : 1.57
5 : 2.02

Figure 6.10 Contour map of the truncated surface as visualized in Figure 6.9.

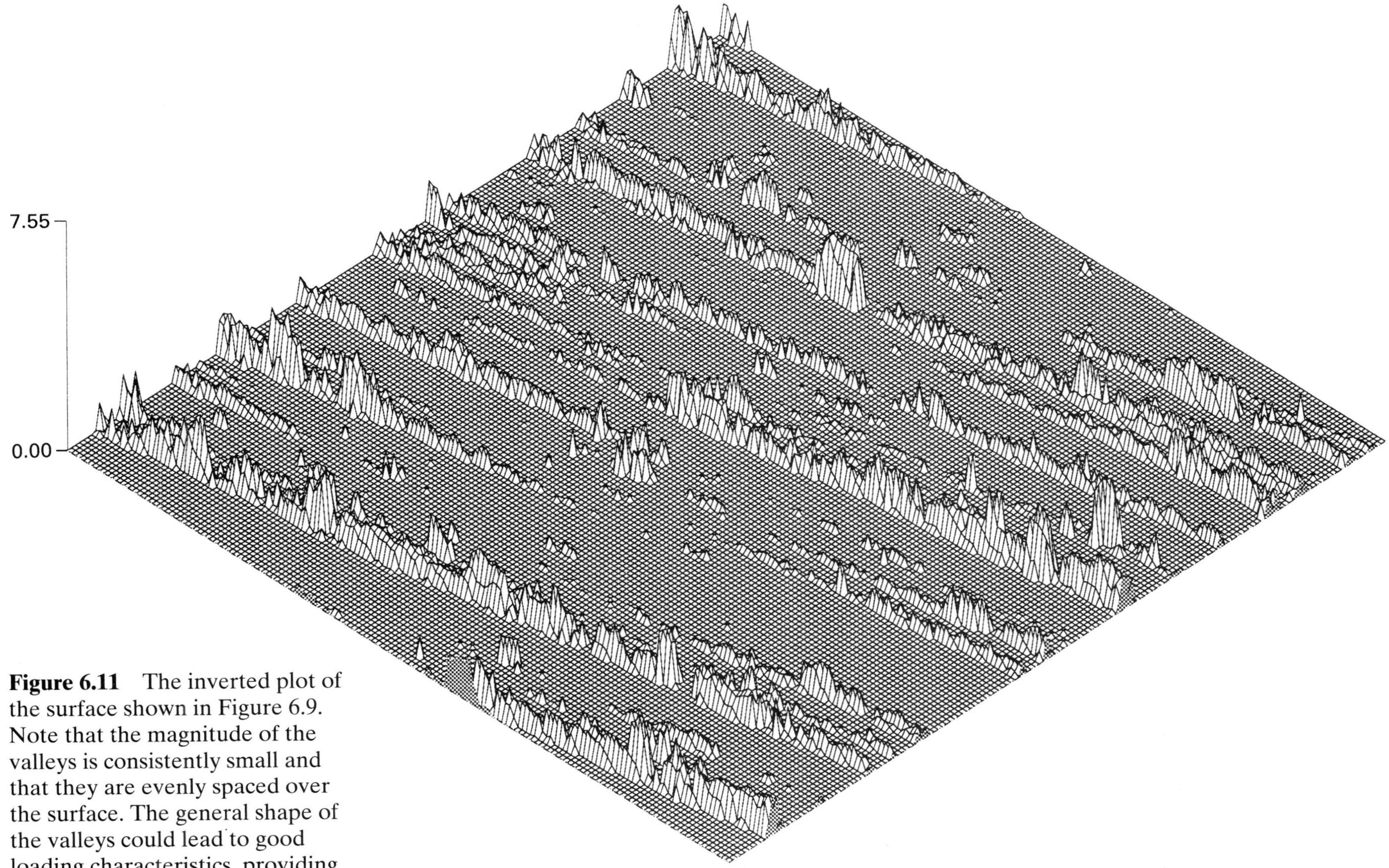

Figure 6.11 The inverted plot of the surface shown in Figure 6.9. Note that the magnitude of the valleys is consistently small and that they are evenly spaced over the surface. The general shape of the valleys could lead to good loading characteristics, providing the surface with good tribological properties.

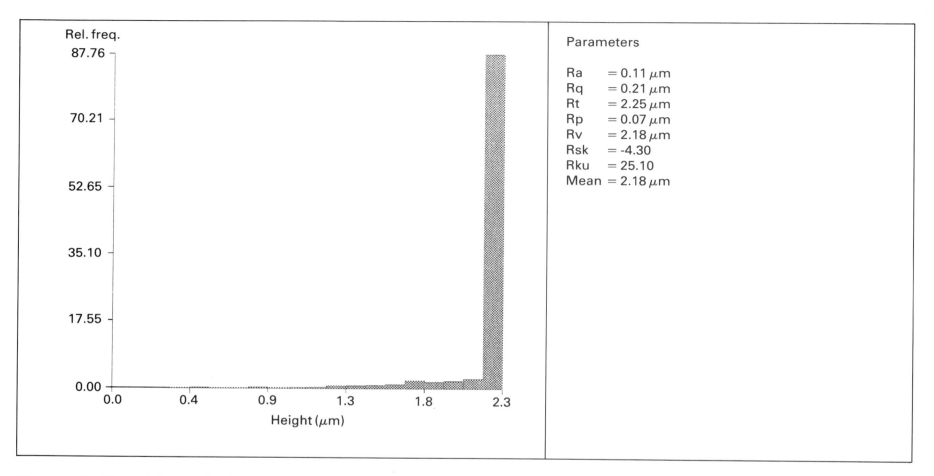

Figure 6.12 The height distribution and associated parameters for the 70% *Rt* truncated surface presented in Figure 6.9.

Figure 6.13 (over) A curve of contact % plotted against truncation for the fly-cutting process: note that the contact % at the uppermost level of the surface is very small until 20% *Rt* truncation is reached. At this point the curve slowly starts to increase, becoming approximately linear at 30% *Rt* truncation after which the rate of increase rapidly falls away as the proportion of the surface in the valleys decreases, becoming approximately zero at the 87% *Rt* truncation level.

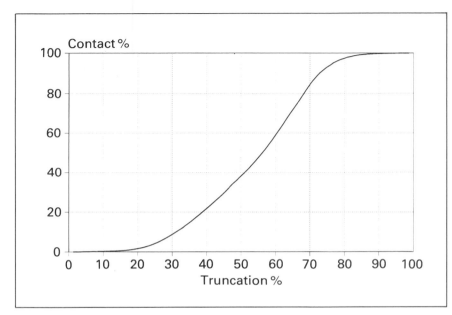

Figure 6.13

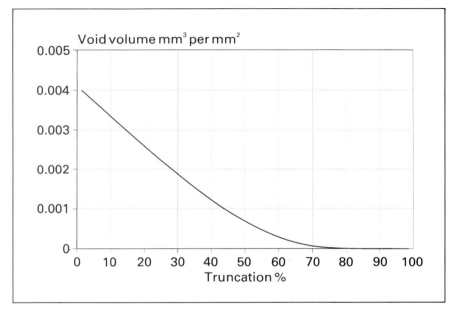

Figure 6.14

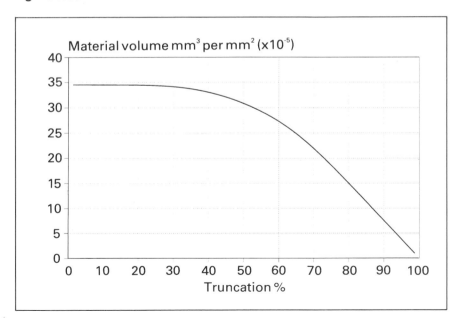

Figure 6.15

Figure 6.14   This graph shows the decrease in void volume as the surface is truncated. Note that the reduction in void volume is approximately linear with truncation level until the 55% *Rt* truncation level is reached. Beyond this level of truncation the void volume decreases more rapidly until it ceases to have any practical value at the 70% *Rt* truncation level. This is a typical feature of surfaces produced using a conventional single-point cutting tool.

Figure 6.15   A graph of material volume remaining in the asperities plotted against the truncation %. As is clearly indicated in the figure, the material in the asperities does not substantially reduce until the 30% *Rt* truncation level is reached. From 30% *Rt* truncation the rate of material removal steadily increases, reaching a maximum rate at approximately 60% *Rt* truncation level. This rate continues until all the asperities are removed. This rate of removal is constant due to the fact that almost all the material being removed below the 60% level is bulk material; the remaining valley contributes very little material saving from the bulk surface.

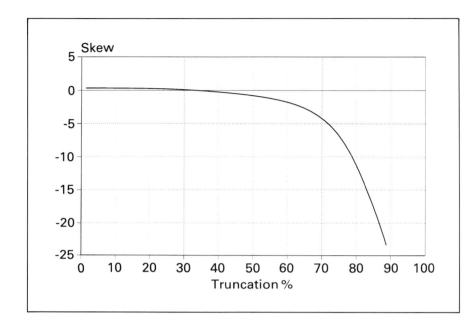

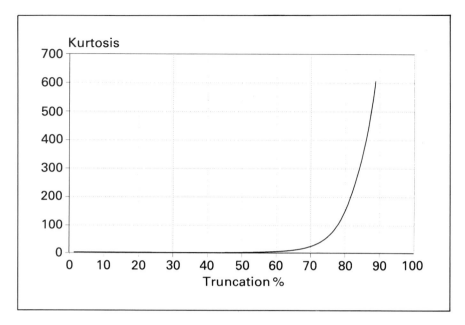

Figure 6.16 As stated previously, the skewness parameter is slightly positive in the original machined condition. As truncation proceeds the skewness parameter decreases slightly until the rate of increase becomes more significant at the 50% *Rt* truncation level.

Figure 6.17 The kurtosis parameter remains constant at a value of approximately *Rku* = 3.0 until it rapidly starts to increase at 60% *Rt* truncation level. After this point it rapidly approaches infinity.

7 BORED SURFACES

The surface described in this section is typical of those generated by the boring process. As Figure 7.1 shows, the surface displays a number of segments or ridges parallel to each other caused by the action of a single-point cutting tool as it cuts successive grooves into the surface. The distance between each groove is related to the feed-rate of the boring tool. The tools used in these processes are typically round-nosed and this feature is easily detected from the figure. Notice that the height of the walls of each ridge on the surface is determined, and limited, by the relationship between the depth of cut of the cutting tool as it passes through the bore and the feed-rate per revolution of the cutter. The uneven height of the walls is primarily caused by variations in the cutting forces which arise at the tool-chip interface as a result of the non-homogeneity of the material which is being cut. The resulting surface can be seen to have a random structure in one direction but it is a highly structured (correlated) trend in the opposing direction.

An interesting feature that only becomes evident when the surface is inverted (Figure 7.11) is that the tool tip appears to have a defect causing a ridge towards the centre of the radius. This ridge repeats itself on each curvature. If the ridge only showed itself as one groove (inverted) it could be attributed to adhered particles on the tool face, but the fact that it repeats so consistently means that it must be caused by a tool tip defect. Inversion of the asperities has again revealed features which are not always obvious in the normal 3-D representation.

The shape of the surface is described most effectively in terms of the shape parameters, skewness and kurtosis parameters, which have values typical of processes which employ a single-point cutting tool. Single-point cutting processes yield positive skewness and low values of kurtosis.

As has already been stated in relation to other surfaces, the kurtosis value is mathematically related to the skewness value. The trends shown in this case are typical of those found with surfaces generated by a single-point action and the convenient spacing between valleys leads to good tribological performance. Such a surface has good potential as a tribological surface, but to achieve the best results attention should be paid to the cutting tool geometry and feed-rate.

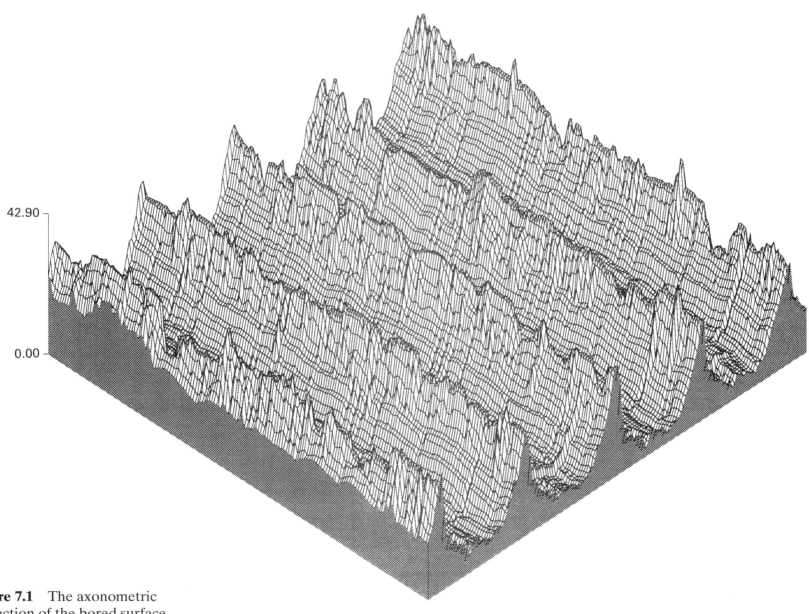

42.90 —

0.00 —

Figure 7.1 The axonometric
projection of the bored surface.

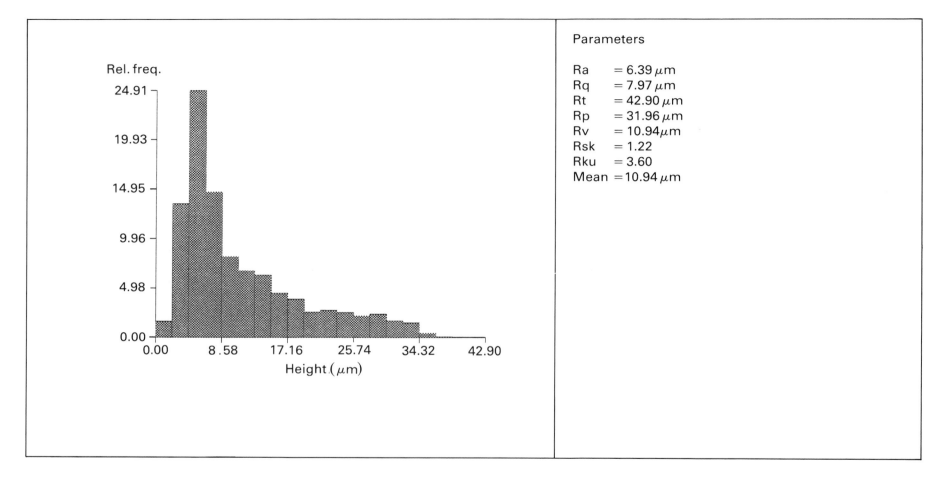

Figure 7.2 The distribution of heights is biased towards the valleys, a feature which is obvious from Figure 7.1. At the higher levels the frequency of asperities decreases exponentially. The resulting parameters show that the surface is very rough ($Ra = 6.39$ μm; $Rt = 42.90$ μm).

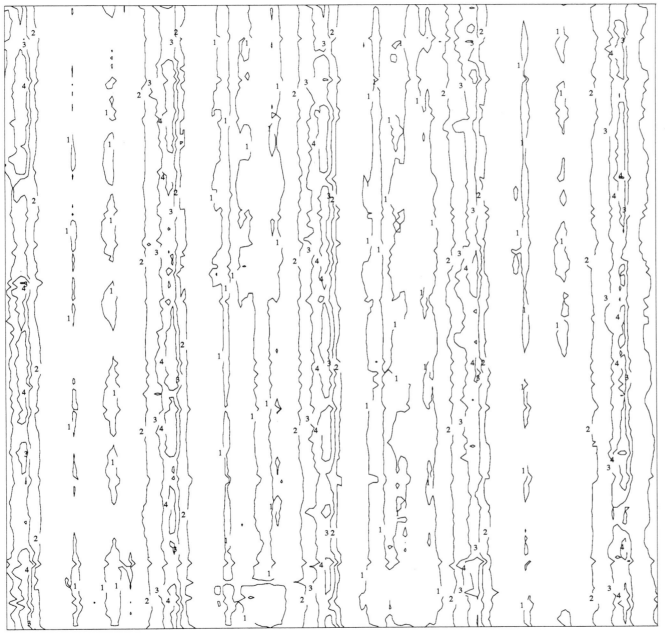

Contour key (μm)

1 : 4.29	4 : 30.03
2 : 12.87	5 : 38.61
3 : 21.45	

Figure 7.3 The contour map shows the structured nature of the surface very clearly.

Figure 7.4 (over) These four diagrams present composite information on successive traces taken from the surface in two orthogonal directions. As would be expected from the projection shown in Figure 7.1 the various parameters, Ra', Rq' and skewness, of the surface differ significantly. The Ra' and Ra' values are closely similar in trend, as indicated in the two diagrams, and both show that the roughness parameters are small along the lay whilst being quite large across the lay (the direction in which most 2-D assessments are taken). The skewness parameters are quite different in the two directions. In this example the two skewness curves are similar in shape but offset. The standard deviations of the two curves are similar. The differences in the kurtosis values are relatively small, as indicated in the diagram.

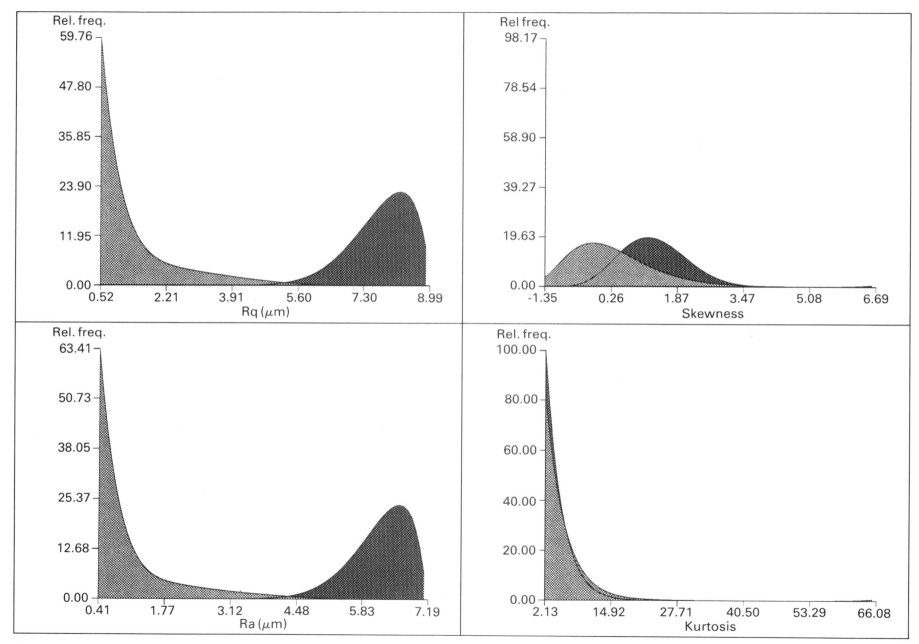

Figure 7.4

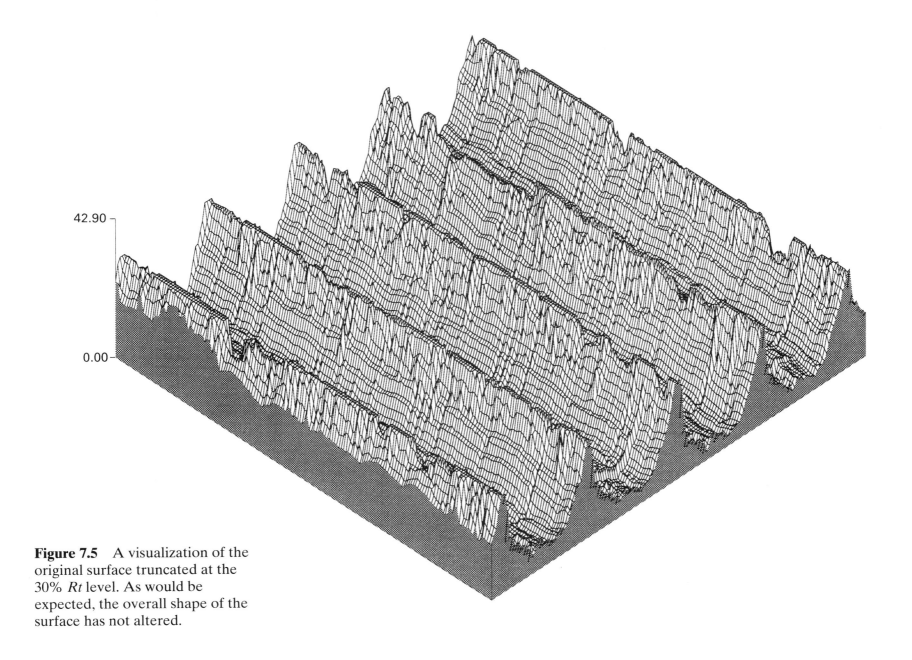

42.90

0.00

Figure 7.5 A visualization of the original surface truncated at the 30% *Rt* level. As would be expected, the overall shape of the surface has not altered.

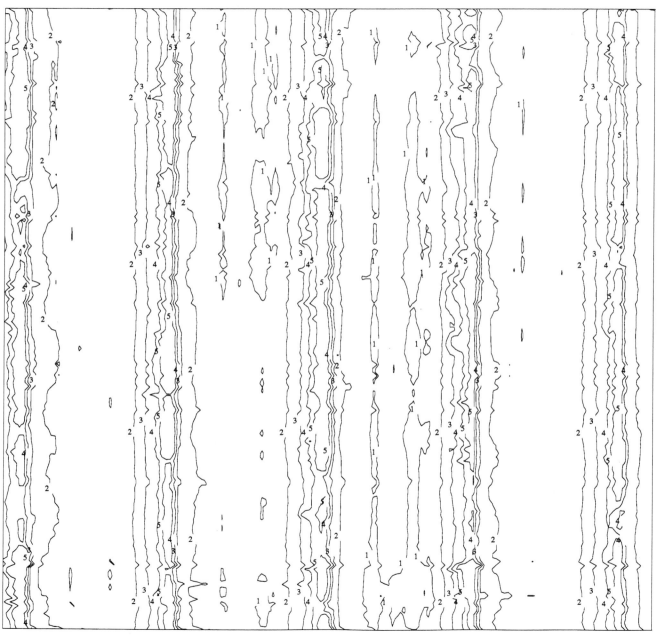

Contour key (μm)

1 : 3.00
2 : 9.00
3 : 15.00
4 : 21.00
5 : 27.00

Figure 7.6 The contour map of the truncated surface, like the projection shown in Figure 7.5, differs little from that of the original surface.

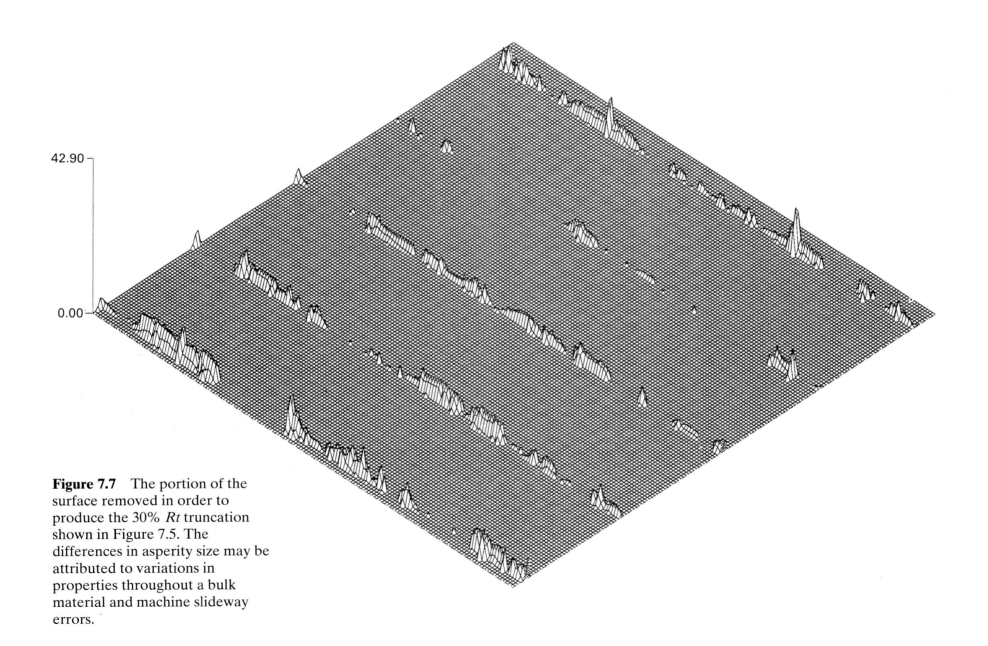

42.90

0.00

Figure 7.7 The portion of the surface removed in order to produce the 30% *Rt* truncation shown in Figure 7.5. The differences in asperity size may be attributed to variations in properties throughout a bulk material and machine slideway errors.

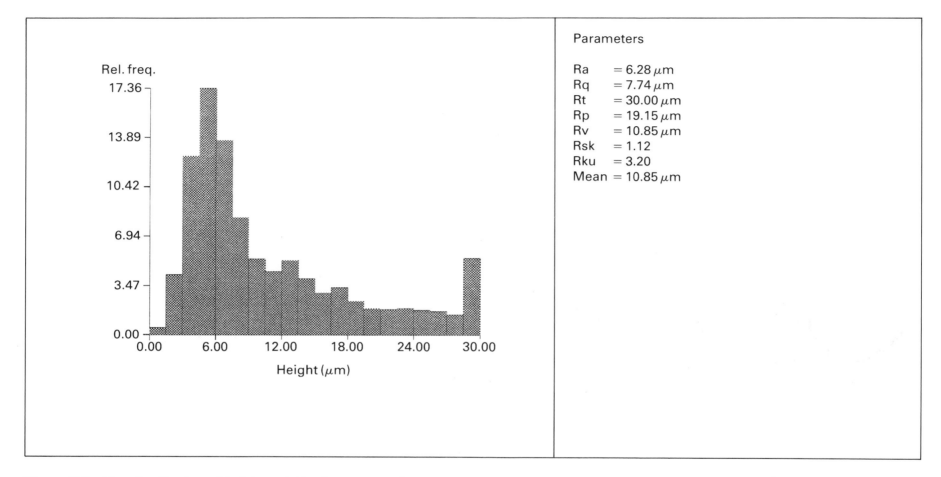

Parameters

Ra	= 6.28 μm
Rq	= 7.74 μm
Rt	= 30.00 μm
Rp	= 19.15 μm
Rv	= 10.85 μm
Rsk	= 1.12
Rku	= 3.20
Mean	= 10.85 μm

Figure 7.8 The distribution of heights within the truncated surface: the right hand end of the distribution curve is forming an impulse as a result of truncation and, as a consequence, the area skewness parameter and the area kurtosis parameters are reducing; a feature which is always found as positively skewed surfaces are truncated.

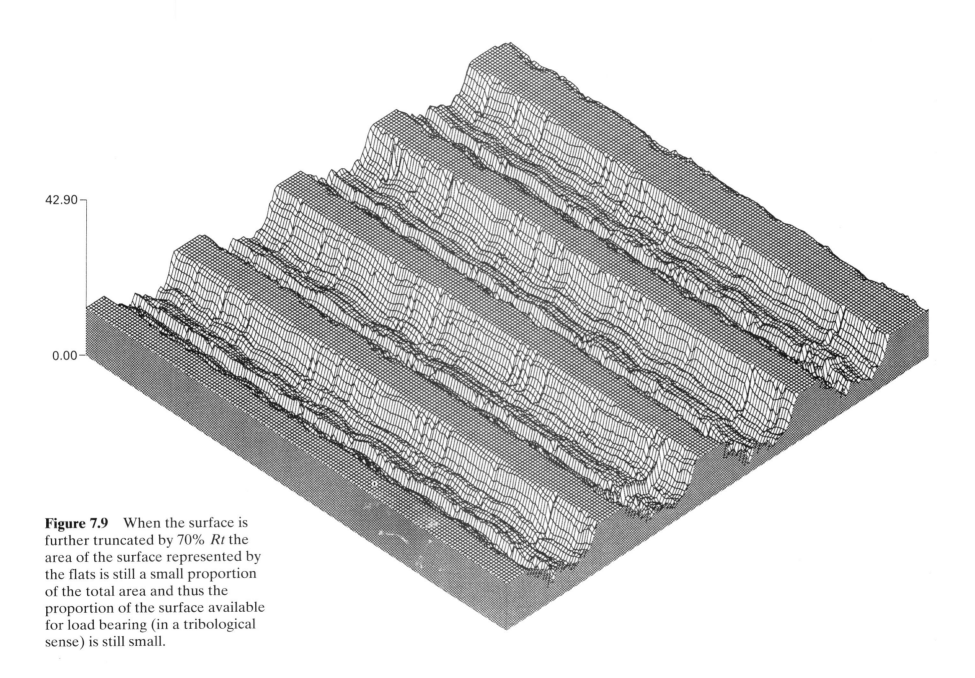

42.90 —

0.00 —

Figure 7.9 When the surface is further truncated by 70% *Rt* the area of the surface represented by the flats is still a small proportion of the total area and thus the proportion of the surface available for load bearing (in a tribological sense) is still small.

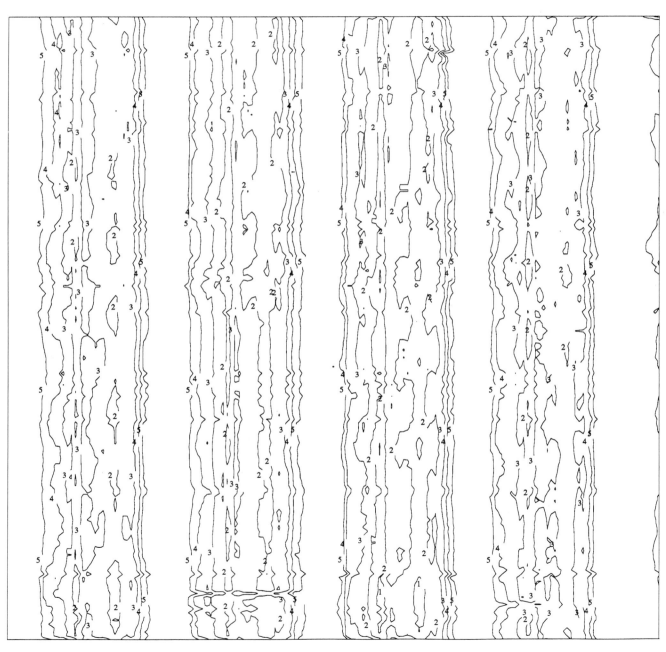

Contour key (μm)

1 : 1.29
2 : 3.87
3 : 6.45
4 : 9.03
5 : 11.61

Figure 7.10 The contour plot reinforces the point that only a small proportion of the surface is available for load-bearing even at high truncation levels.

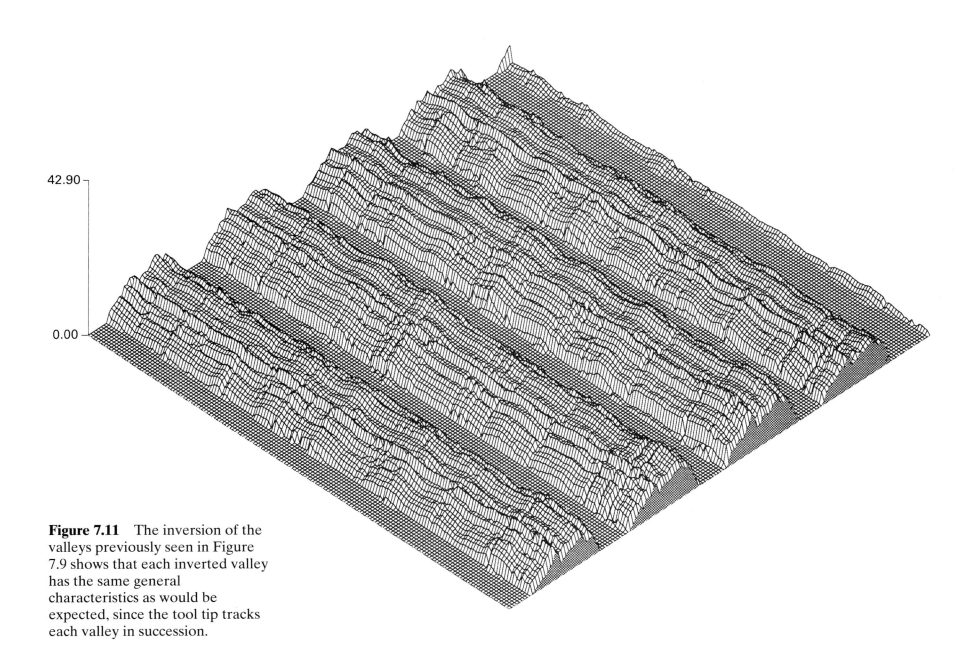

42.90

0.00

Figure 7.11 The inversion of the valleys previously seen in Figure 7.9 shows that each inverted valley has the same general characteristics as would be expected, since the tool tip tracks each valley in succession.

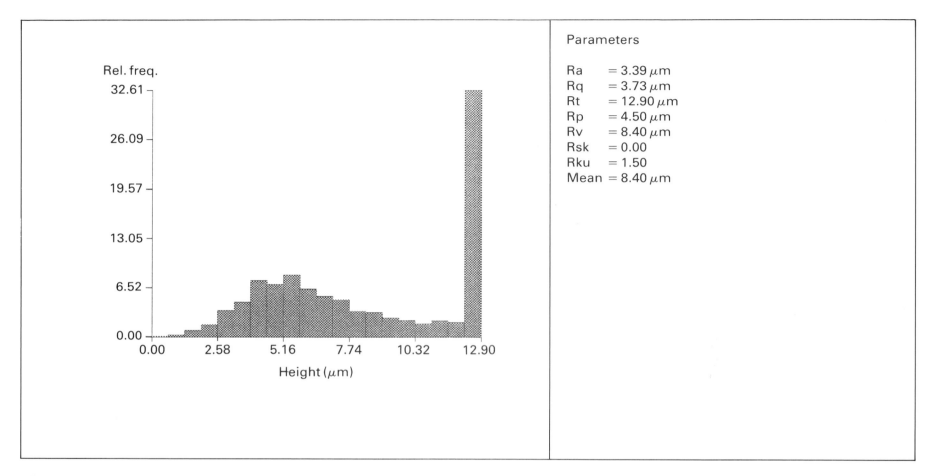

Parameters

Ra   = 3.39 μm
Rq   = 3.73 μm
Rt   = 12.90 μm
Rp   = 4.50 μm
Rv   = 8.40 μm
Rsk = 0.00
Rku = 1.50
Mean = 8.40 μm

Figure 7.12   The 3-D height parameters of the surface truncated to 70% show that the assembly of the resulting profile is still marginal. The shape parameters indicate a particularly low value of kurtosis (*Rku* = 1.50).

Figure 7.13 (over)   This graph shows a plot of contact % against truncation % for the bored surface. The nature of the curve on the graph indicates that there is a very slow transition from low contact % to high contact %. This trend is similar to that found for turned surfaces as shown in Section 1.

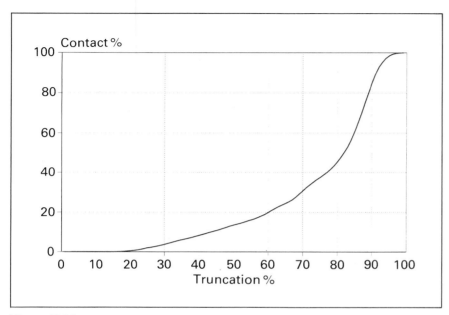

Figure 7.13

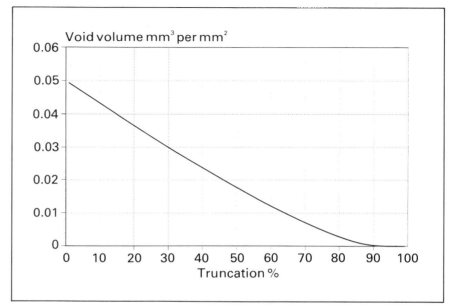

Figure 7.14

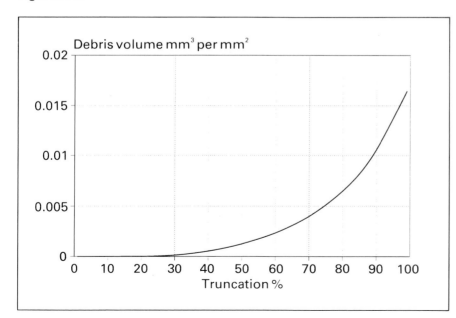

Figure 7.15

Figure 7.14   The curve for void volume against truncation % shows that, in this case, the void volume, which represents the volume above the lowest asperity, reduces rapidly until the 82% Rt truncation level is reached. Note that the void volume is reduced to 10% at approximately the 72% Rt truncation level. This low value for void volume is attributed to the shape of the grooves in the surface which relates to the shape of the point of the boring tool.

Figure 7.15   A graph of debris volume plotted against truncation %. As would be expected, and as has been shown in earlier diagrams in other sections, the debris which is generated at early levels of truncation is very small. The rate of increase is hardly detectable until the 30% Rt truncation level is reached. From that point onwards the rate of removal steadily increases until the 80% Rt truncation level is reached and then increases linearly with further truncation. From the 80% Rt level the bulk of material removed comes from the body of the material since very little is lost due to the presence of the rapidly diminishing valleys.

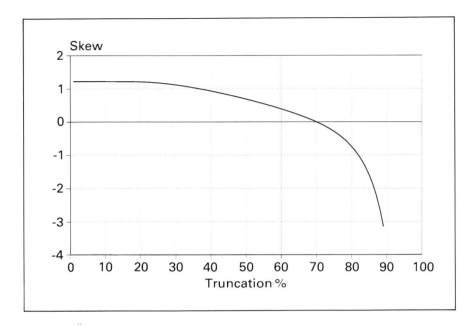

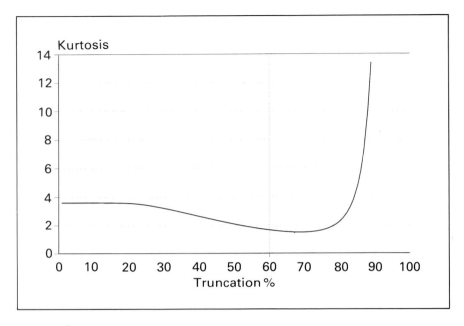

Figure 7.16   The plot of the skewness parameter as it develops with truncation shows that the surface is initially positively skewed (*Rsk* = +1.2) which is typical of a surface which has been generated using a single-point cutting tool. As truncation proceeds, the surface reaches zero skewness at approximately 70% *Rt* truncation and then rapidly proceeds to negative skewness, reaching a value of –3.0 at the 88% *Rt* truncation level.

Figure 7.17   The kurtosis value is approximately 3.5 at zero truncation, reducing towards the value of 3.0 at the 30% *Rt* truncation level. The value continues to reduce, reaching a minimum value of 1.5 at approximately 70% *Rt* truncation. From this point it rises rapidly, reaching the value of 12 at 88% *Rt* truncation value.

8 SLAB MILLED SURFACES

The figures presented in this section relate to a surface produced by the slab milling process. The surface has a highly correlated structure, generated in this case by a combination of the feed-rate and the peripheral velocity of the milling cutter during the cutting process. Where there are departures from correlation of the surface structure this is due to the random action at the tool-chip interface, caused in part by the non-homogeneity within the material composition. This causes variations in the cutting force, together with variations in coolant supply to the tool-chip interface and finally wear of the cutting edge itself. The magnitude of the average roughness value, Ra, can vary dramatically during slab milling, and is related to feed-rate per revolution of the cutting teeth as the surface moves under the cutter, and also to the depth of cut per tooth.

An interesting feature of the sample surface is the indication of a slight waviness in the feed direction of the workpiece. This is caused, most probably, by small form errors in the machine tool slideways and machine spindle rotational inaccuracies. This feature can be seen to repeat itself across the entire width of the projection (Figure 8.1) which would be expected from a geometric cause of origin as suggested.

As is evident from Figure 8.16, the asymmetry parameter, skewness, is most unusual in that it does not follow the steady curve seen in previous examples, but indicates different trends over various sections of the curve. Such effects can only be explained by the existence of waviness on the surface.

Even when the surface is subjected to simulated wear (Figures 8.5 and 8.6) the valleys which still exist enable the feed per tooth to be calculated if required. It is interesting to note that the errors associated with spindle deflection are more significant than the errors associated with slideways effects. Hence there are indications that better surface quality could be achieved by increasing the size and stiffness of the milling machine cutter spindle.

In conclusion, this particular surface would be unsuitable for tribological applications, and this is due not only to the underlying process but also to the condition of the machine on which the surface was generated. This set of figures indicates how topography can be used as a diagnostic tool in machine-health monitoring.

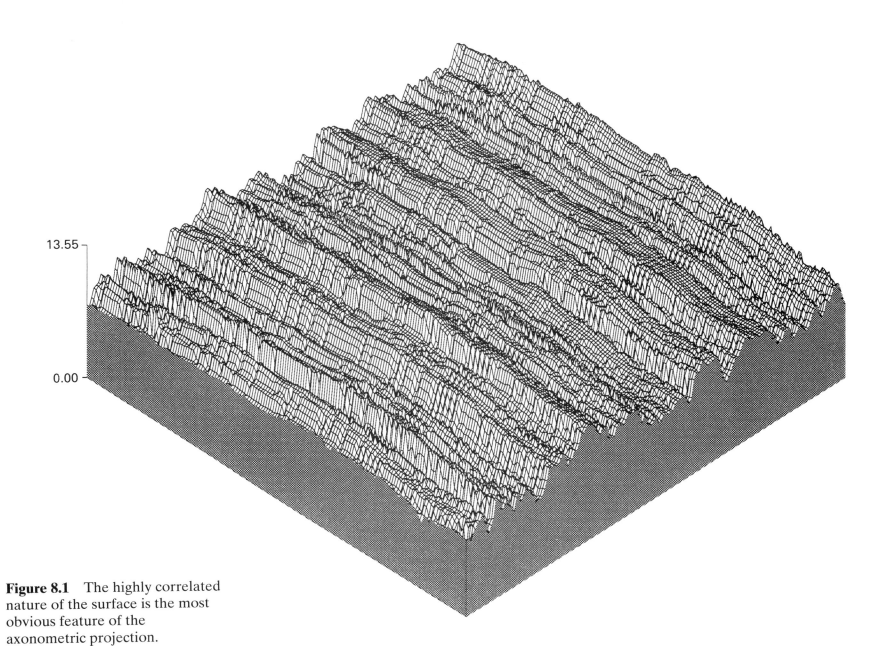

13.55

0.00

Figure 8.1 The highly correlated nature of the surface is the most obvious feature of the axonometric projection.

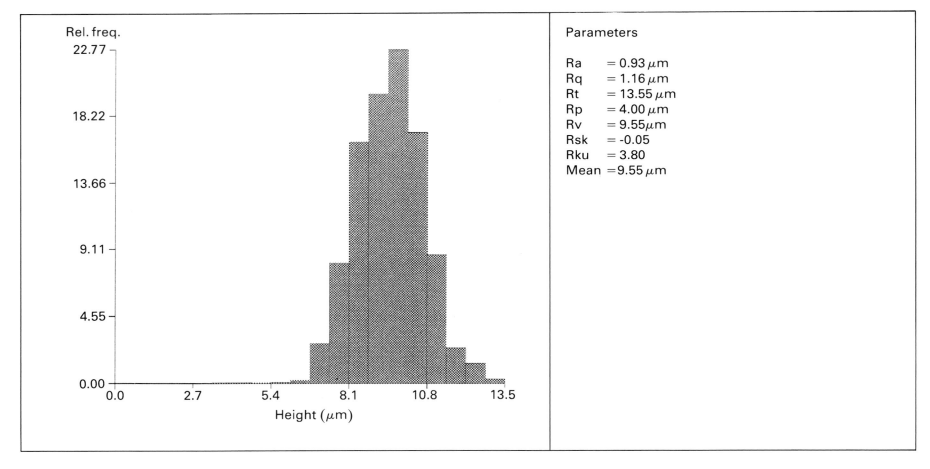

Figure 8.2 The histogram of the asperity heights for the 3-D
surface shows that the distribution of asperities is generally
Gaussian. This distribution is slightly offset due to the occurrence
of a small number of events, namely the occasional deep valley
randomly positioned within the surface which slightly affects the
area parameter values which have been computed for the surface.
The Gaussian nature of the surface is indicated numerically by the
values of skewness and kurtosis which are typical of the slab milling
process.

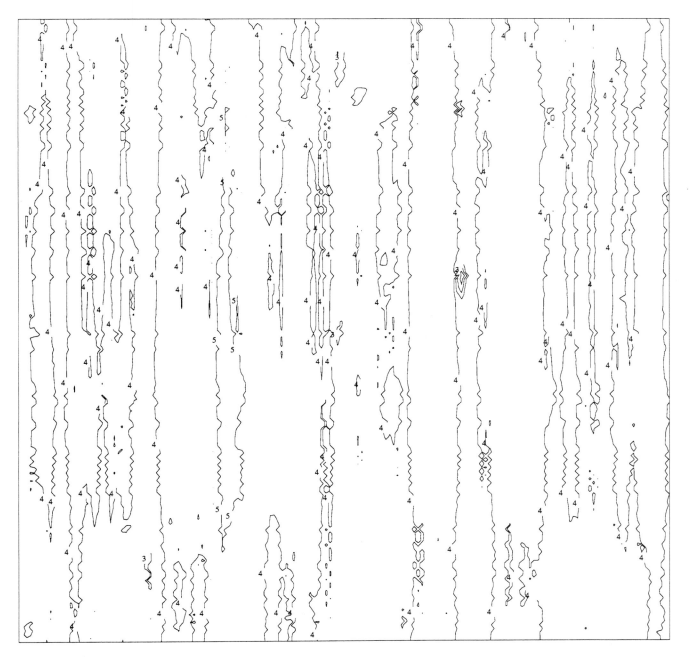

Contour key (μm)

1 : 1.35
2 : 4.06
3 : 6.77
4 : 9.48
5 : 12.19

Figure 8.3 The contour map shows the directional nature of the cutting process. This feature is interrupted only by the effects of micro-roughness caused by the factors discussed in the text.

Figure 8.4 (over) These four curves represent Ra', Rq', skewness and kurtosis of the individual traces taken in orthogonal directions across the surface. For the slab milled surface shown, the four curves are significantly different. The roughness parameters, Ra' and Rq', both indicate that measurements across the lay yield larger values than those along the lay. The shape parameter, skewness, also indicates that measurements across the lay yield higher values than those along the lay, although the differences are not as great as would have been expected from a visual interpretation of the surface. Kurtosis values, by contrast, are closely similar.

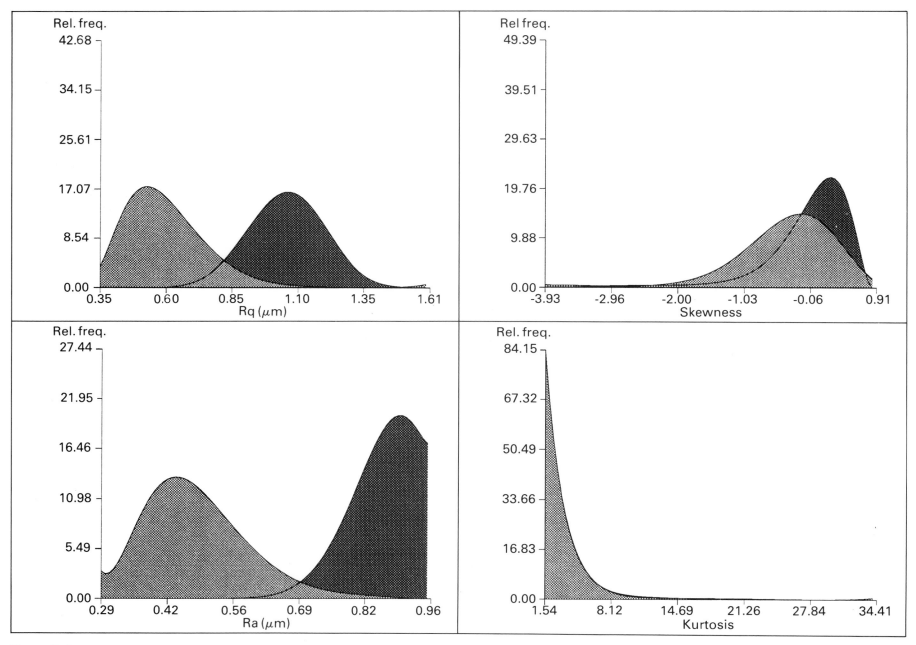

Figure 8.4

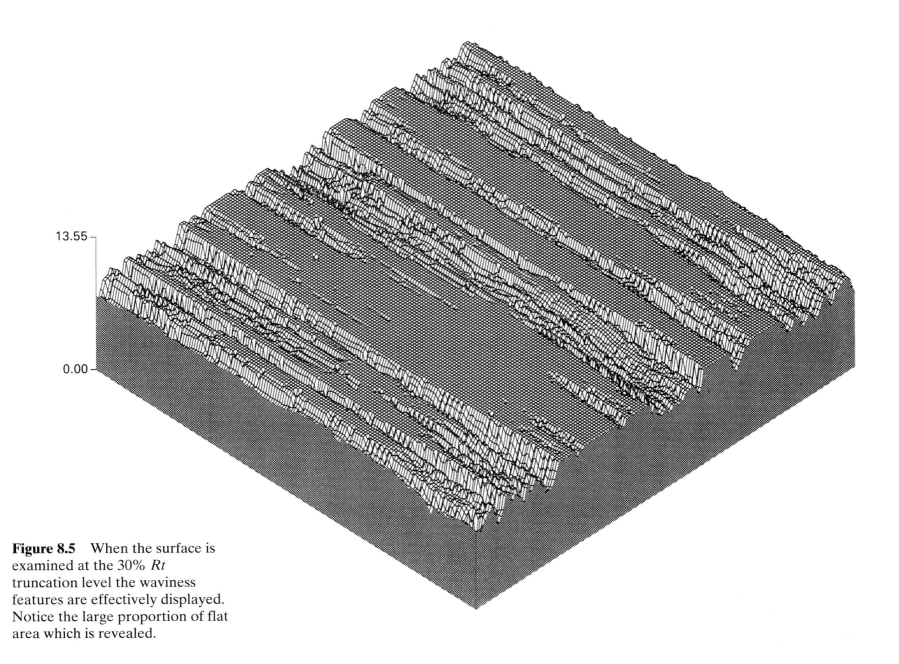

13.55 —

0.00 —

Figure 8.5 When the surface is examined at the 30% *Rt* truncation level the waviness features are effectively displayed. Notice the large proportion of flat area which is revealed.

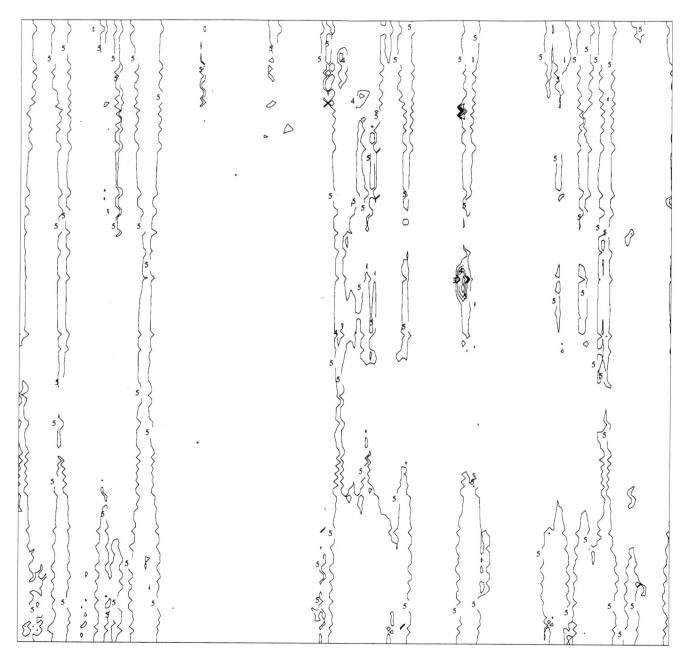

Contour key (μm)

1 : 0.95
2 : 2.85
3 : 4.75
4 : 6.65
5 : 8.55

Figure 8.6 The contour map of the surface after 30% Rt truncation shows that large structural changes have occurred to the surface, and this diagram complements the observations made concerning Figure 8.5.

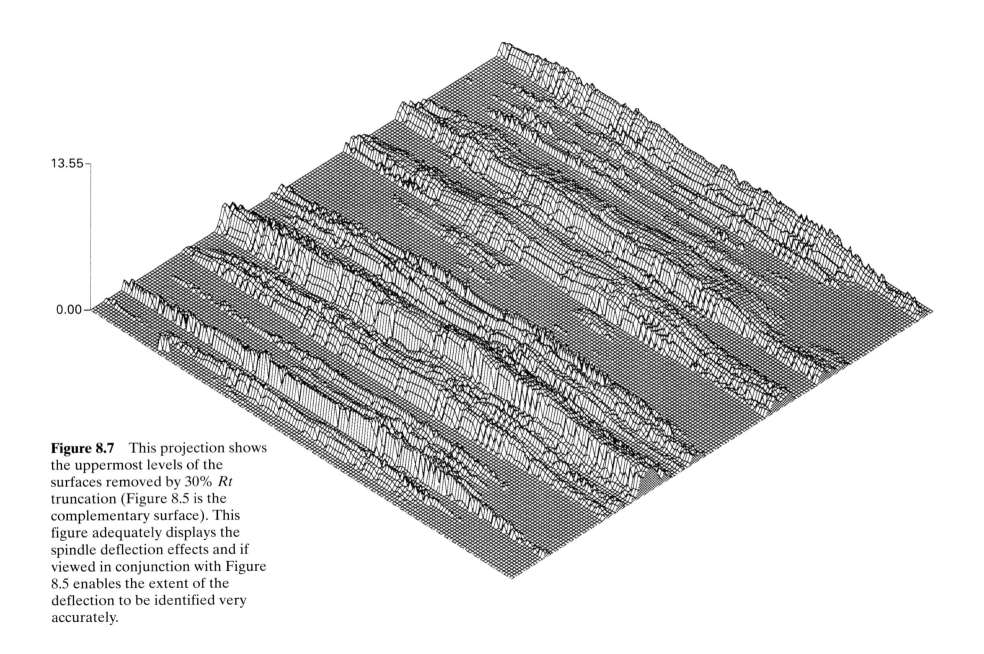

13.55 —

0.00 —

Figure 8.7 This projection shows the uppermost levels of the surfaces removed by 30% *Rt* truncation (Figure 8.5 is the complementary surface). This figure adequately displays the spindle deflection effects and if viewed in conjunction with Figure 8.5 enables the extent of the deflection to be identified very accurately.

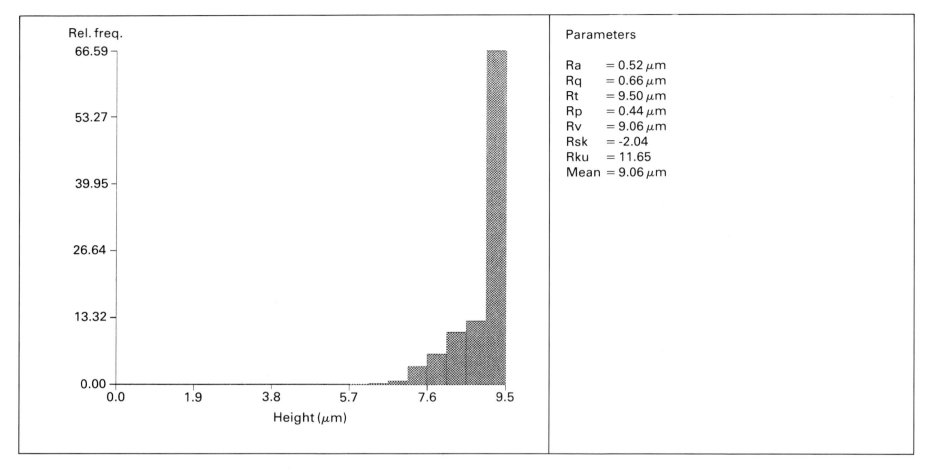

Figure 8.8 This figure represents the distribution of asperity heights for the truncated surface shown in Figure 8.5. Note that the impulse on the histogram represents the removed portion of the curve. The surface is becoming heavily skewed, but due to the distribution of valleys is likely to have poor load-bearing qualities. The complementary kurtosis value is high.

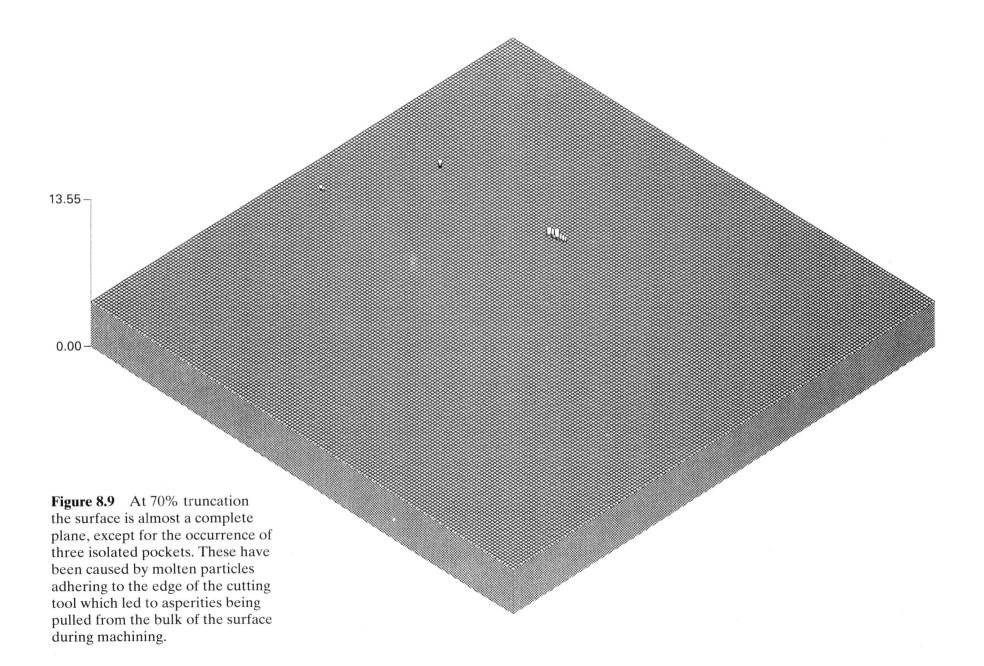

13.55

0.00

Figure 8.9 At 70% truncation the surface is almost a complete plane, except for the occurrence of three isolated pockets. These have been caused by molten particles adhering to the edge of the cutting tool which led to asperities being pulled from the bulk of the surface during machining.

Contour key (μm)

1 : 0.40
2 : 1.21
3 : 2.02
4 : 2.83
5 : 3.64

Figure 8.10 The contour map of the 70% *Rt* truncated surface shown in Figure 8.9.

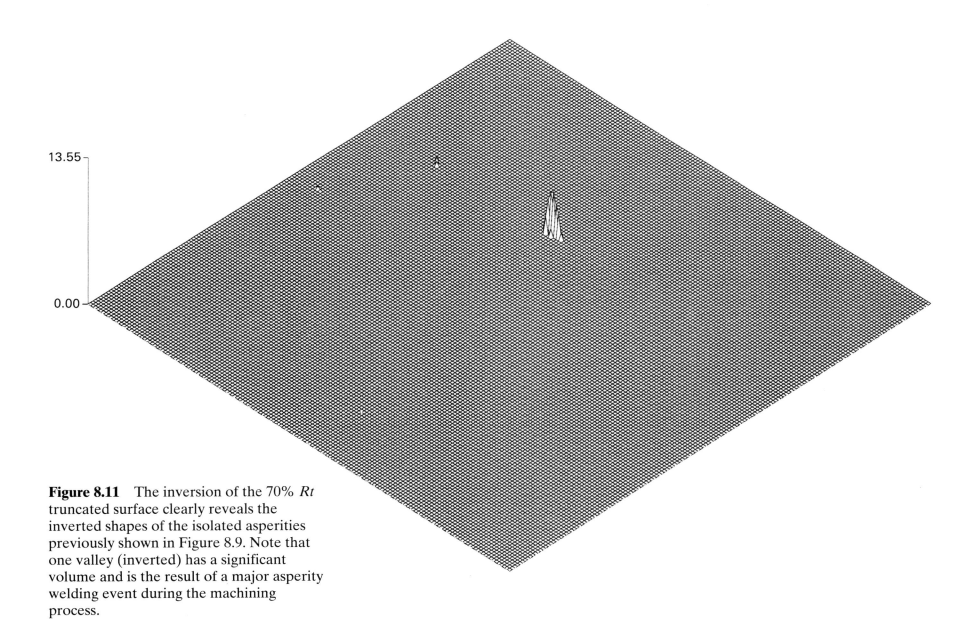

13.55

0.00

Figure 8.11 The inversion of the 70% *Rt* truncated surface clearly reveals the inverted shapes of the isolated asperities previously shown in Figure 8.9. Note that one valley (inverted) has a significant volume and is the result of a major asperity welding event during the machining process.

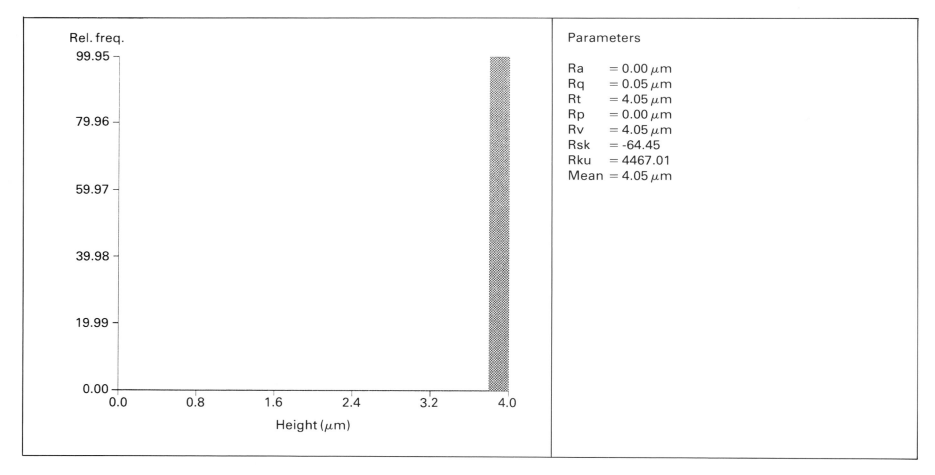

Figure 8.12 The height distribution for the surface after 70% *Rt* truncation indicates that all the weighting occurs due to the uppermost level. Very little is attributed to the remaining pockets. Skewness and kurtosis parameters have no significance at these values, but it can be visually observed that the surface would not lubricate well.

Figure 8.13 (over) A graph of contact % against truncation % plotted for the slab milled surface shows that at early stages of truncation very little contact is achieved and this is largely due to the waviness feature which was evident in Figure 8.1. At 15% *Rt* truncation the surface contact % starts to increase and reaches a maximum rate at approximately 22% *Rt* truncation. This trend continues until approximately the 45% *Rt* truncation level, when the bulk of the remaining voids in the surface are either due to the waviness feature previously described, or due to the occurrence of the few catastrophic asperity events mentioned earlier.

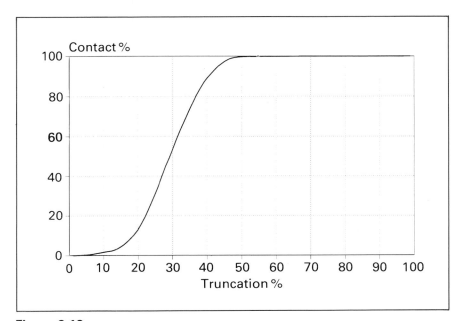

Figure 8.13

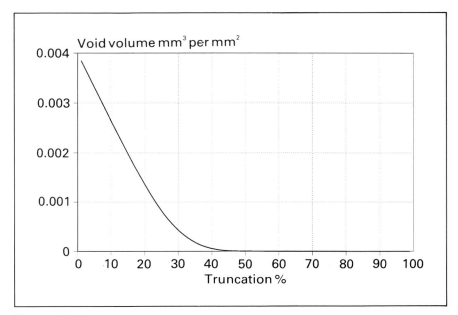

Figure 8.14

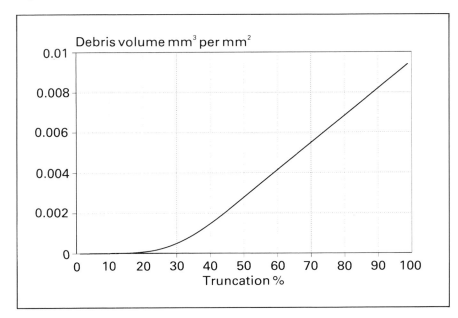

Figure 8.15

Figure 8.14   The void volume parameter shows a trend complementary to that demonstrated in Figure 8.13. The reduction in void volume is constant until the 28% Rt truncation level is reached. After this point the void volume rapidly reduces, and by 40% Rt truncation few voids exist to provide a reservoir for lubrication. Those that do remain are largely confined to the asperity catastrophes previously described.

Figure 8.15   Until 20% Rt truncation is reached very little debris is formed, and that which is produced comes purely from the upper levels of the micro-asperities. As the 20% Rt truncation level is passed the waviness features are becoming significant and debris is being formed from these at 35% Rt truncation. From this point onwards the majority of the debris is coming from the bulk material and hence the debris curve remains approximately constant throughout the remainder of the truncation process.

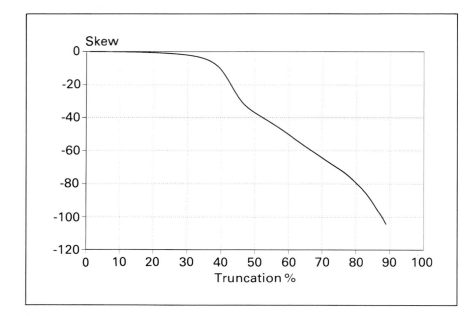

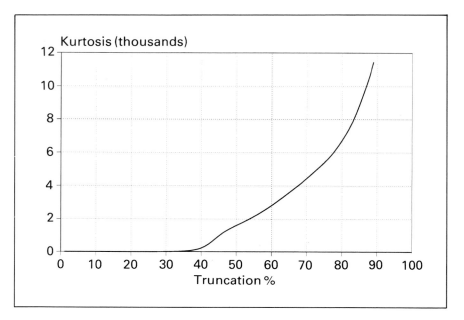

Figure 8.16 The skewness parameter is most unusual in that the curve indicates different trends at various levels of truncation. The first turning point indicates the onset of the influence of waviness and the second turning point indicates that the bulk properties of the surface are starting to have an overriding effect.

Figure 8.17 The kurtosis parameter indicates trends similar to those of the skewness curve for similar reasons. The kurtosis values seen in the figure are so large as to mask much of the meaning that can be attributed to the parameter.

9 SHAPED SURFACES

The figures presented in this section relate to a typical surface produced by the shaping process.

The distance between the tips of the asperities, which can be seen in Figure 9.1, relates to the feed-rate of the tool per cutting pass. Notice that the sides of the valleys are reasonably consistent, but there is evidence of 'twin valleys' due to an imperfection on the tool tip which affected the shearing action between the tool and work material as the surface was removed. The projection indicates that the waviness present in the surface is related only to the tool-point feed-rate relationship and not to any form errors caused by machine tool induced errors. The micro-roughness which is present is a function of the tool-chip interaction, and indicates a smooth and efficient metal removal process.

The combination of skewness and kurtosis parameters shown in Figure 9.2 is typical of surfaces produced by single-point cutting tools, and has been seen with previous processes demonstrated in this atlas. The average roughness value Ra is 2.65 μm and the magnitude of this parameter is purely related to the tool shape and feed-rate per pass of the cutting tool.

The roughness across the lay, however, is significantly greater than that along the lay (Figure 9.4) and it is therefore important, if this surface is to be investigated in 2-D, that the measurements be made across the lay if the maximum value of roughness is to be determined. The skewness parameter, although showing different distributions, is not so severely affected by the direction of measurement. The kurtosis parameter shows larger differences, with the across-lay values having the greater magnitude.

The contour map of the original surface (Figure 9.3) suggests that the surface will not change in terms of orientation as truncation proceeds, and subsequent figures confirm this. Note that the valleys which have been formed would provide useful oil retention features if the surface was used in a tribological environment. The conclusion can be drawn that the combination of cutting tool tip and tool-chip interaction produces a very consistent form on the machined surface.

In general the shaped surface, like the turned surface, has a good potential in a tribological application. The well distributed valleys could promote long term running life by providing a number of sources for lubrication retention to promote hydrodynamic lubrication and to prevent surface-to-surface contact.

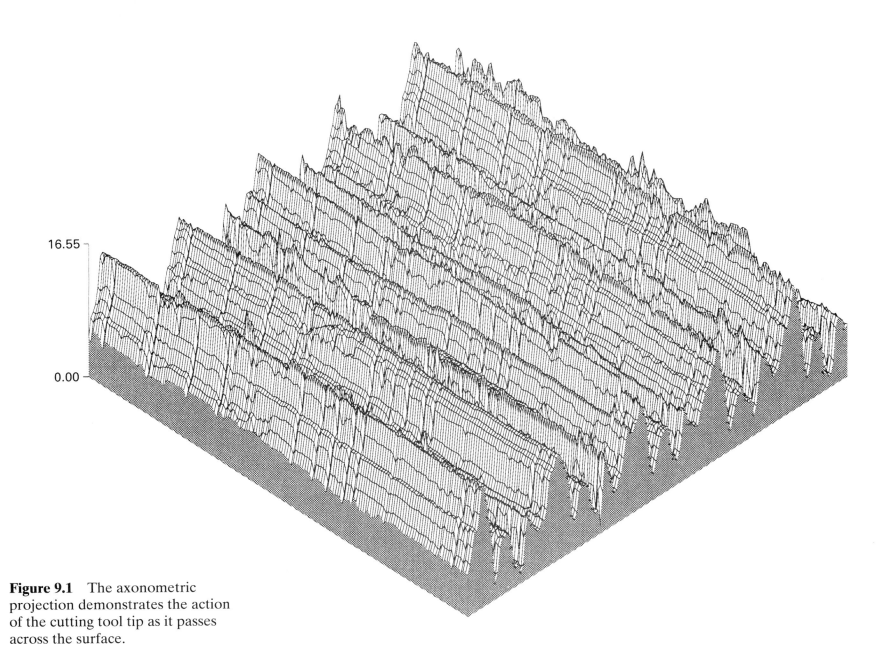

16.55

0.00

Figure 9.1 The axonometric projection demonstrates the action of the cutting tool tip as it passes across the surface.

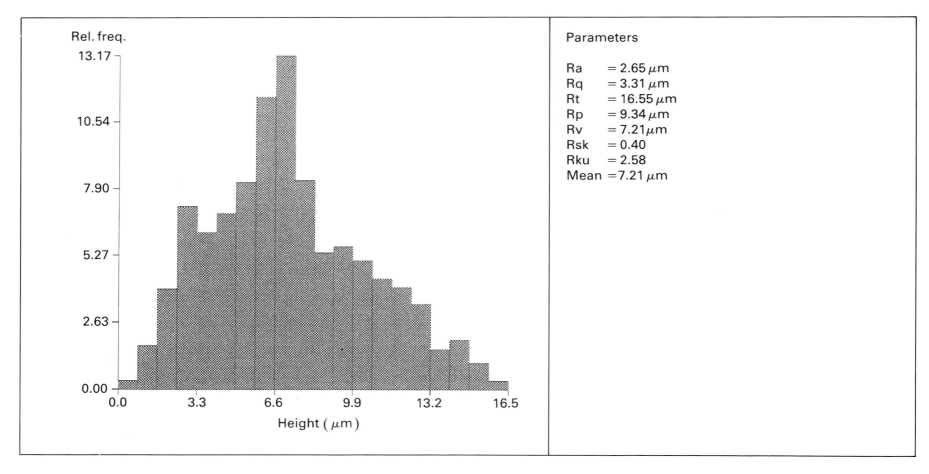

Figure 9.2 The histogram of the asperities within the surface shows that the surface is positively skewed and has a low kurtosis value.

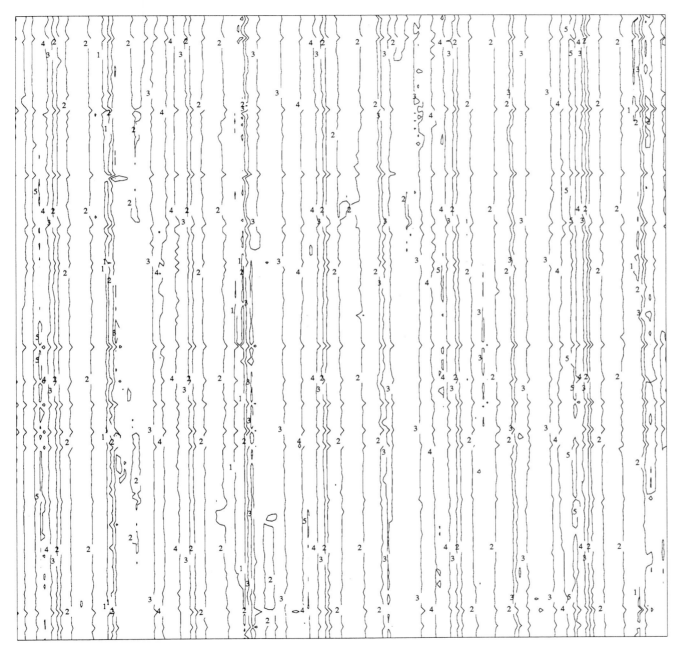

Contour key (μm)

1 : 1.65
2 : 4.96
3 : 8.27
4 : 11.58
5 : 14.89

Figure 9.3 The contour map indicates the unidirectional nature of the cutting process; this feature is so strongly present that it should remain dominant through all truncation levels examined.

Figure 9.4 (over) Curves of the parameters Ra' and Rq' show closely similar results, and both indicate quite clearly that the roughness of the shaped surface across the lay is significantly different to the roughness along the lay.

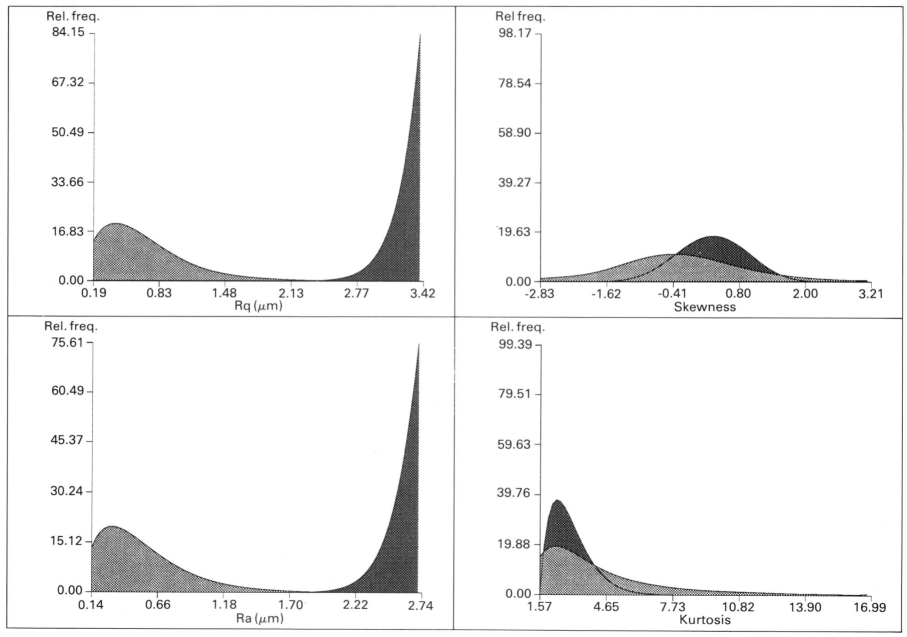

Figure 9.4

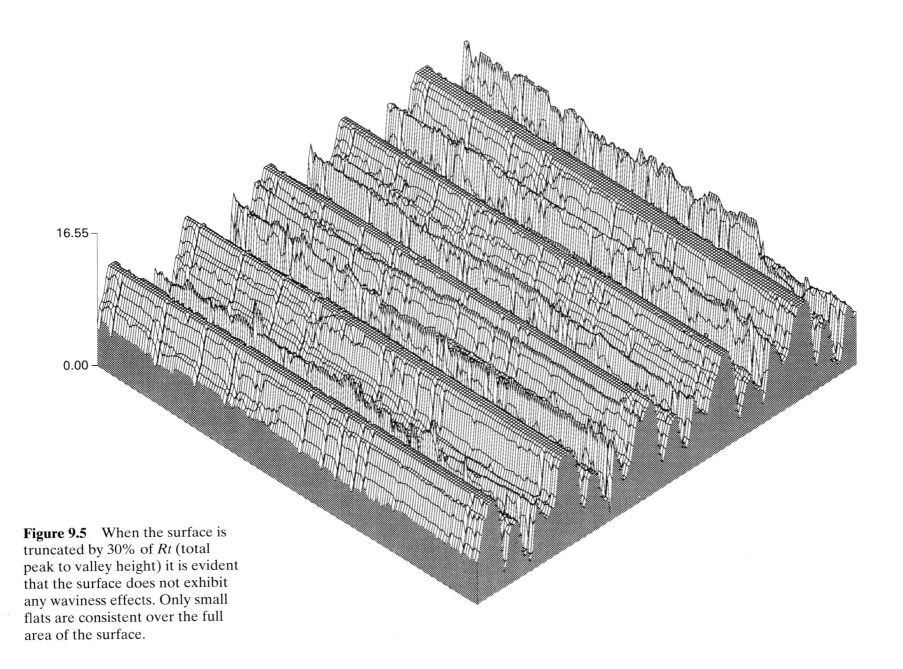

16.55

0.00

Figure 9.5 When the surface is truncated by 30% of *Rt* (total peak to valley height) it is evident that the surface does not exhibit any waviness effects. Only small flats are consistent over the full area of the surface.

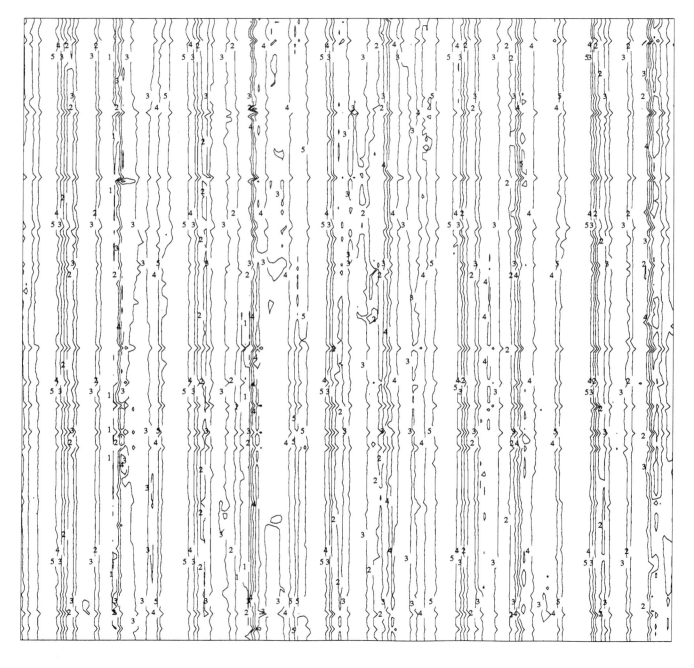

Contour key (μm)

1 : 1.16
2 : 3.48
3 : 5.80
4 : 8.12
5 : 10.44

Figure 9.6 The contour map of the surface after 30% *Rt* truncation has not significantly changed with truncation, and is indistinguishable from the previous plot shown in Figure 9.3.

16.55 —

0.00 —

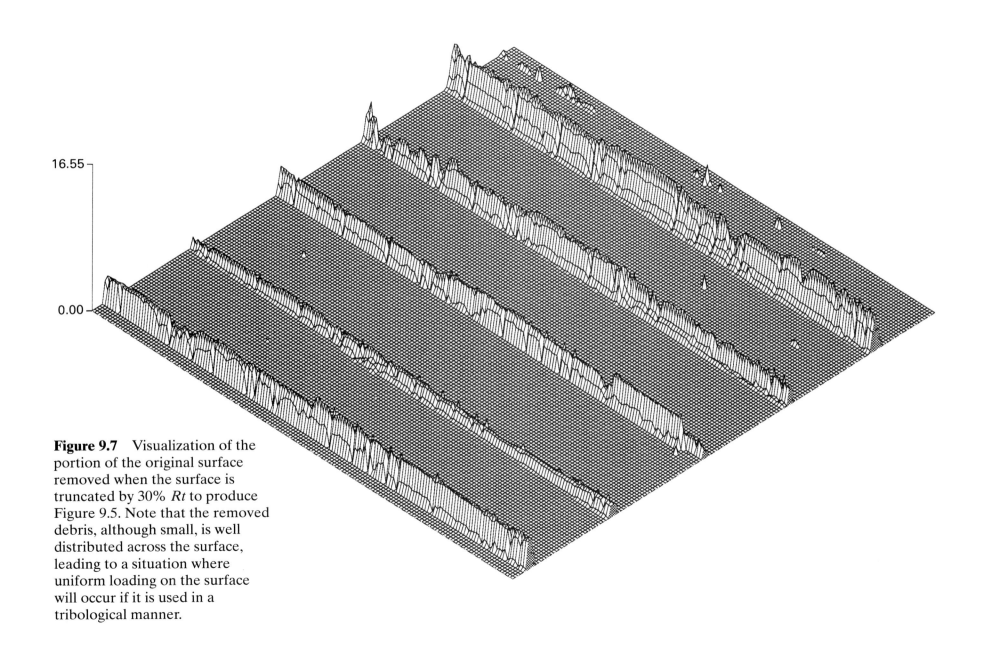

Figure 9.7 Visualization of the portion of the original surface removed when the surface is truncated by 30% *Rt* to produce Figure 9.5. Note that the removed debris, although small, is well distributed across the surface, leading to a situation where uniform loading on the surface will occur if it is used in a tribological manner.

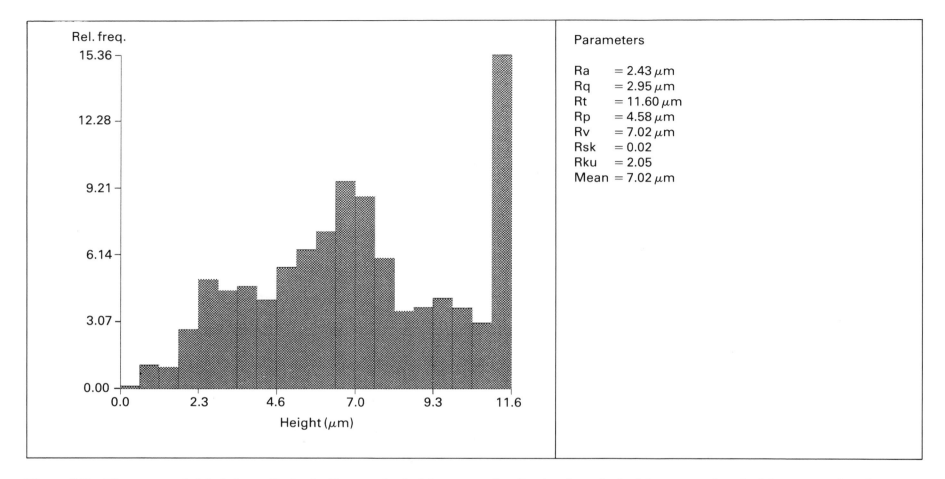

Figure 9.8 The removed debris has a limited effect on the height parameters which describe the surface after 30% *Rt* truncation. Average roughness has been reduced slightly, skewness and kurtosis are also reduced but not significantly. The impulse on the height distribution curve represents the proportion of the surface which is present at the uppermost flat plane. The surface height distribution is typical of those associated with truncated surfaces originally generated by a single-point cutting tool. To demonstrate this point, it is useful to compare the height distribution with that produced for a turned surface, shown in Figure 1.8. Again it is seen that the skewness is positive and the kurtosis has a low value (less than three).

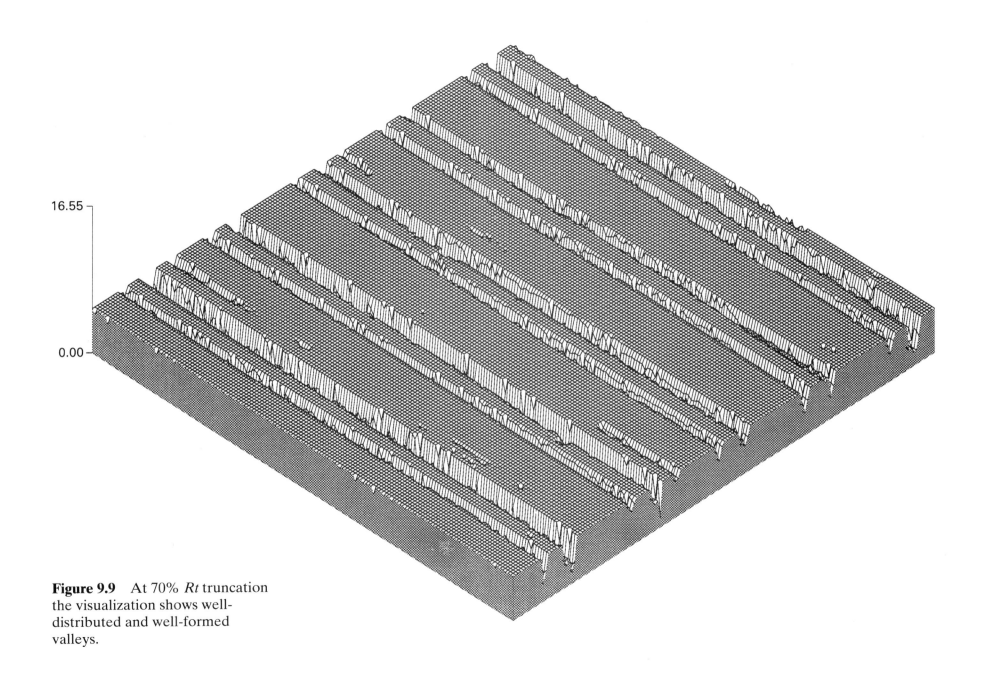

16.55

0.00

Figure 9.9 At 70% *Rt* truncation the visualization shows well-distributed and well-formed valleys.

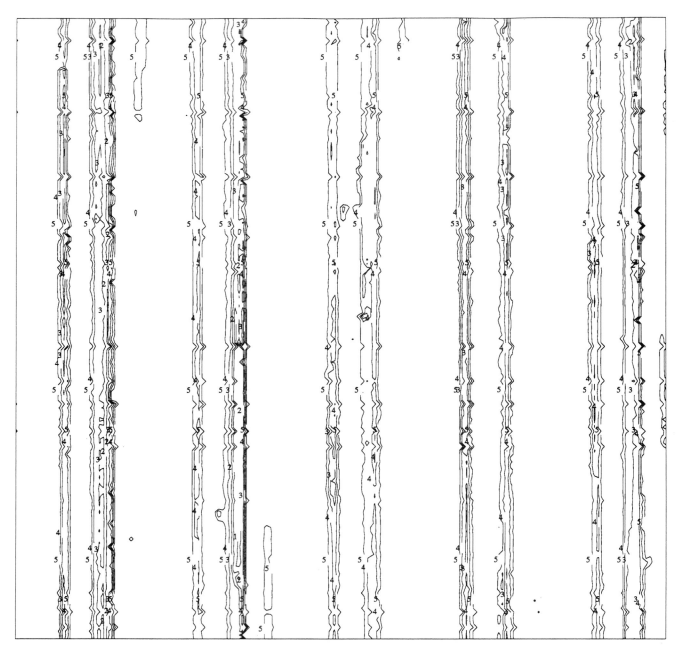

Contour key (μm)

1 : 0.49
2 : 1.48
3 : 2.47
4 : 3.46
5 : 4.45

Figure 9.10 The contour map confirms the even distribution of the valleys evident in Figure 9.9.

16.55 –

0.00 –

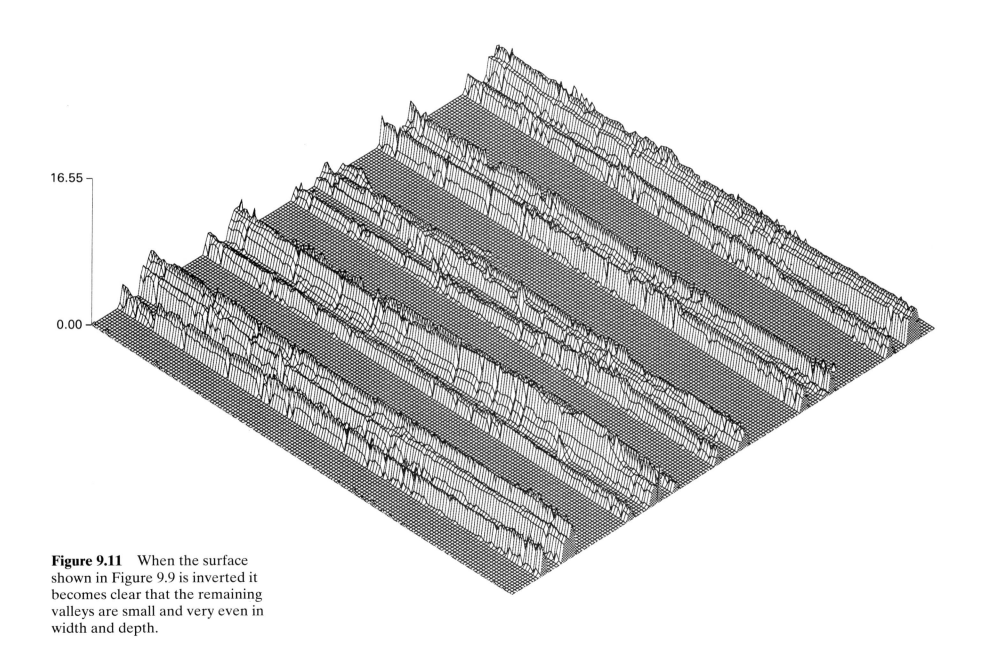

Figure 9.11 When the surface shown in Figure 9.9 is inverted it becomes clear that the remaining valleys are small and very even in width and depth.

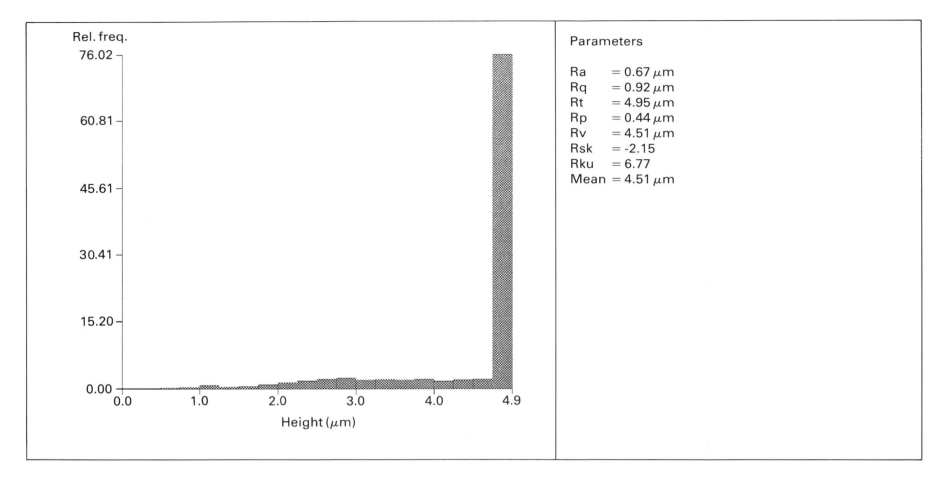

Figure 9.12 The height distribution shows how the remaining surface heights are distributed across the surface after 70% *Rt* truncation. The resulting height distribution parameters, skewness and kurtosis, suggest a surface which would be highly suitable for tribological interactions.

Figure 9.13 (over) Contact % plotted against truncation % for the shaped surface, which has been generated by a single-point cutting process. Again the form of the diagram is very similar to that found previously for the turned surface (Figure 1.8). There is a slow transition from a low contact % to the high contact % which, as reported earlier, is a very useful feature in a tribological surface.

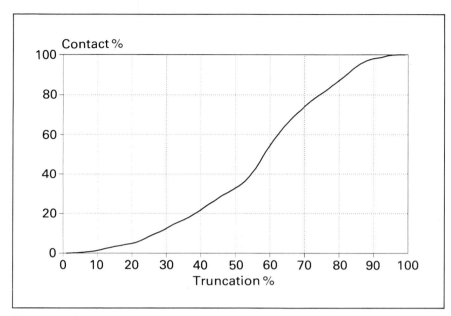

Figure 9.13

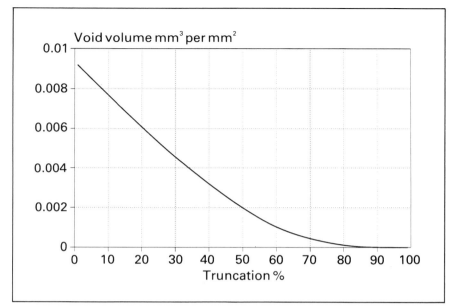

Figure 9.14

Figure 9.14 When the void volume is plotted against truncation % the similarities between the shaped surface and the turned surface are again evident. In this diagram it is shown that the void volume is reduced to 10% of its original value at a truncation value of approximately 62% Rt. This low value is a consequence of a pointed tool tip, and could therefore be modified by changing the tool tip geometry.

Figure 9.15 The volume of debris formed by surface truncation is plotted against truncation %. As can be anticipated from the previous figure, very little debris is removed at the early stages of truncation, and this situation remains relatively unchanged until the 30% Rt truncation level is reached. As the truncation level increases from 30% to 60% Rt, the debris volume increases dramatically, reaching a maximum rate at 70% Rt truncation.

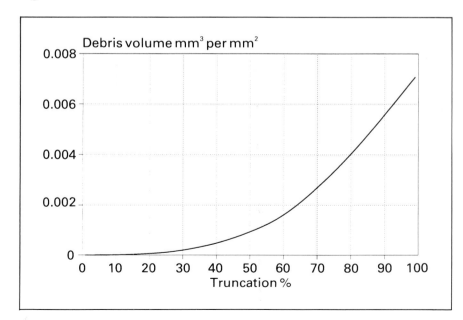

Figure 9.15

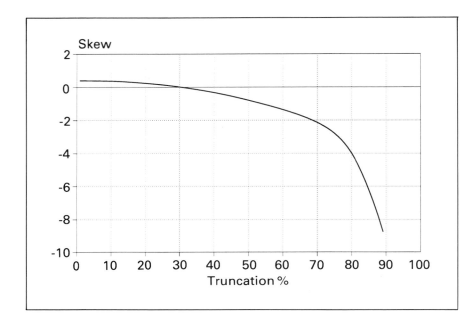

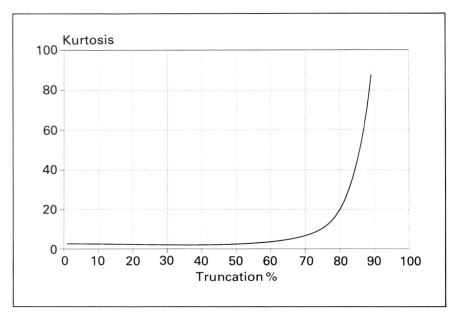

Figure 9.16 The development of the skewness parameter. Prior to truncation the skewness parameter is slightly positive. This parameter reduces to zero at the 30% *Rt* truncation level. At approximately 56% *Rt* truncation the surface is reduced to the mean line (*Rsk* = –1.64). Further truncation leads to the skewness parameter reaching –2.0 at 70% *Rt* and –4.0 at 80% *Rt* truncation (see Figure 1.17 for close similarity). The oil retention grooves are well distributed which could enhance the lubrication properties of the surface.

Figure 9.17 Kurtosis plotted against truncation % shows that until 50% *Rt* truncation is reached, the value of kurtosis does not change significantly. The rate of increase of kurtosis is rapid from this point onwards and at 80% *Rt* truncation it has reached a value of 20. From this point kurtosis increases so rapidly that it becomes meaningless as a controlling surface parameter.

10 POLISHED SURFACES

The figures presented in this section relate to a surface which is typical of that produced by the polishing process. The polishing process is used to produce a flat surface with a high quality of surface finish, i.e. a low Ra value. The axonometric projection (Figure 10.1) shows that such a surface often suffers from loss of form during the polishing process. This loss of form is often associated with an uncontrollable rocking action of the surface during the process. Polishing is produced by the random interaction of a master surface against the modified surface. Since this interaction is often affected by human intervention, errors of form occur. The extent of the form error will vary from component to component. This effect is seen to occur during the production of flat, curved or spherical surfaces. The projection demonstrates that even though polishing is achieved through rotary, or multi-directional actions, the underlying evidence of the previous machining process, in this case fine grinding, is still present. The peak to valley roughness ($Rt = 0.6$ μm), and the average roughness ($Ra = 0.07$ μm), indicate that a very smooth surface has been produced with a form error which is typical of a manual polishing operation. The roughness of such a fine surface could be more effectively analysed using an optical instrument, such as the Wyko or Zygo optical profilometer or indeed the Nanoscope.

The conclusion is that this surface has been badly affected by the generation of a curvature during polishing. This curvature or form error is very common during any polishing process, and is often the largest contributory factor to the total roughness. The form error has been the major contributing factor to numerical values of the parameters of the height distribution. Obviously, this must be an important factor to control during the manufacturing process. This surface indicates some of the limitations of manual polishing when producing components which are required to function with small tolerances.

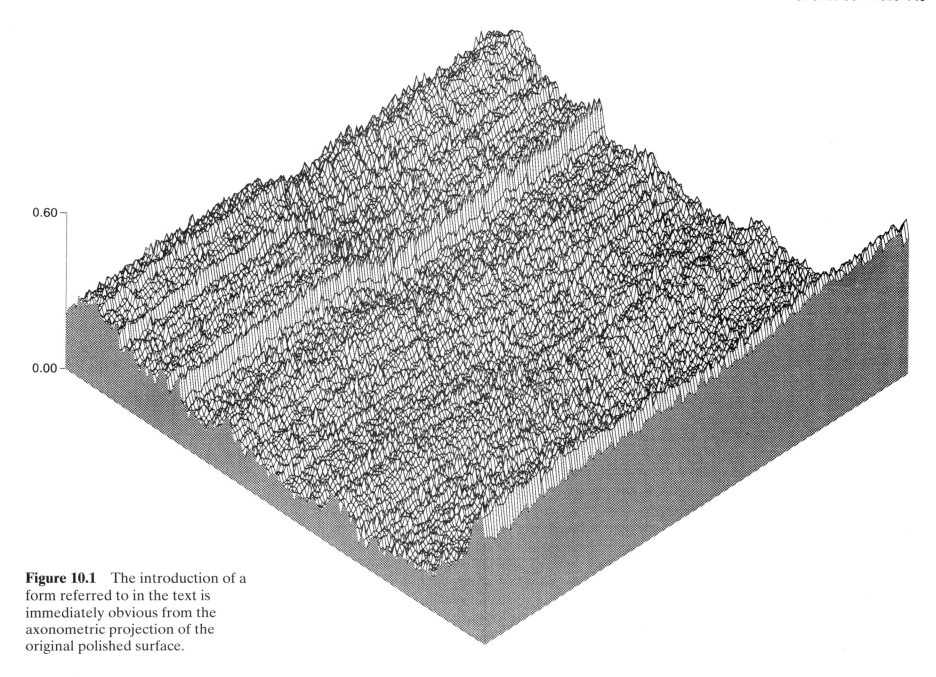

Figure 10.1 The introduction of a form referred to in the text is immediately obvious from the axonometric projection of the original polished surface.

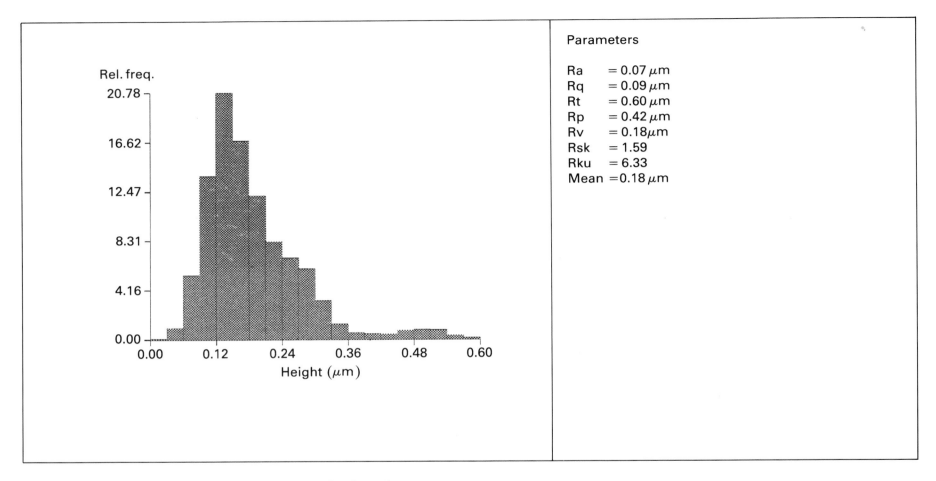

Figure 10.2 The histogram shows that the distribution of
asperities of the polished surface produces a positive skewness
effect indicated by the values of skewness and kurtosis (Rsk = 1.59;
Rku = 6.33). These values are typical of a polished surface.

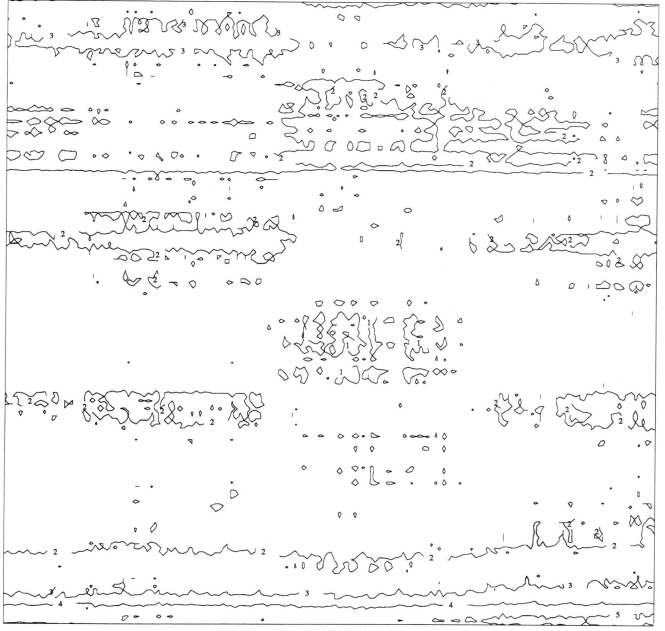

Contour key (μm)

1 : 0.06
2 : 0.18
3 : 0.30
4 : 0.42
5 : 0.54

Figure 10.3 The contour map shows the structural trend of the underlying machining process prior to polishing which is mentioned in the text. Notice that this trend is not regular because polishing has removed some of the original asperities which were produced in the earlier process.

Figure 10.4 (over) Curves of data for Ra', Rq' skewness and kurtosis for 2-D traces taken in orthogonal directions from the area under assessment. The curves show that the roughness parameters of the 2-D profiles are significantly different in the two orthogonal directions. The roughness of the polished surface was significantly affected by the form error generated during the finishing. As a consequence of the form error both the skewness and kurtosis parameters vary significantly. This set of diagrams clearly shows that 2-D assessment is unsuitable for this process.

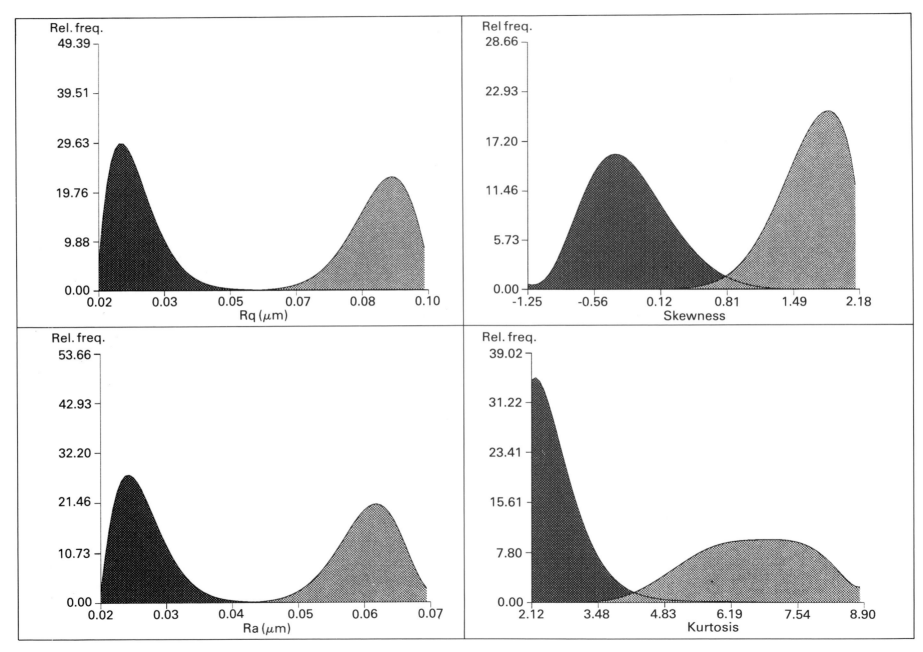

Figure 10.4

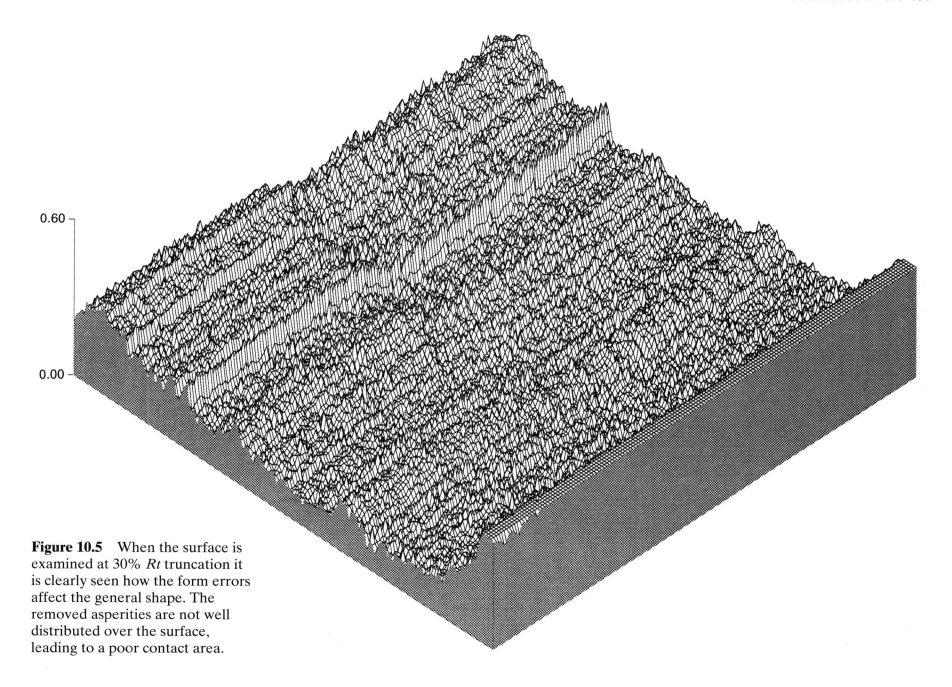

0.60

0.00

Figure 10.5 When the surface is examined at 30% *Rt* truncation it is clearly seen how the form errors affect the general shape. The removed asperities are not well distributed over the surface, leading to a poor contact area.

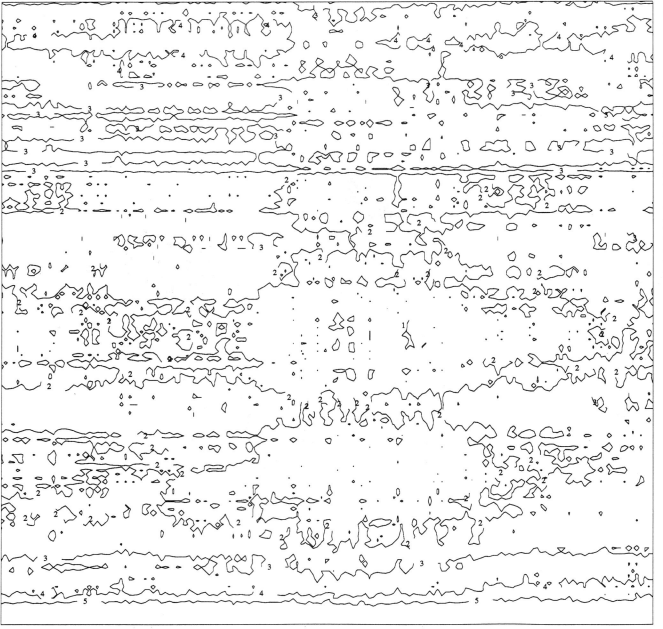

Contour key (μm)

1 : 0.04
2 : 0.13
3 : 0.21
4 : 0.29
5 : 0.38

Figure 10.6 The contour map of the surface shown in Figure 10.5 shows that the truncated surface brings more asperities into consideration indicating that at further truncation more of the surface will be randomly included. Large structural changes are about to be brought into effect when the surface is further truncated.

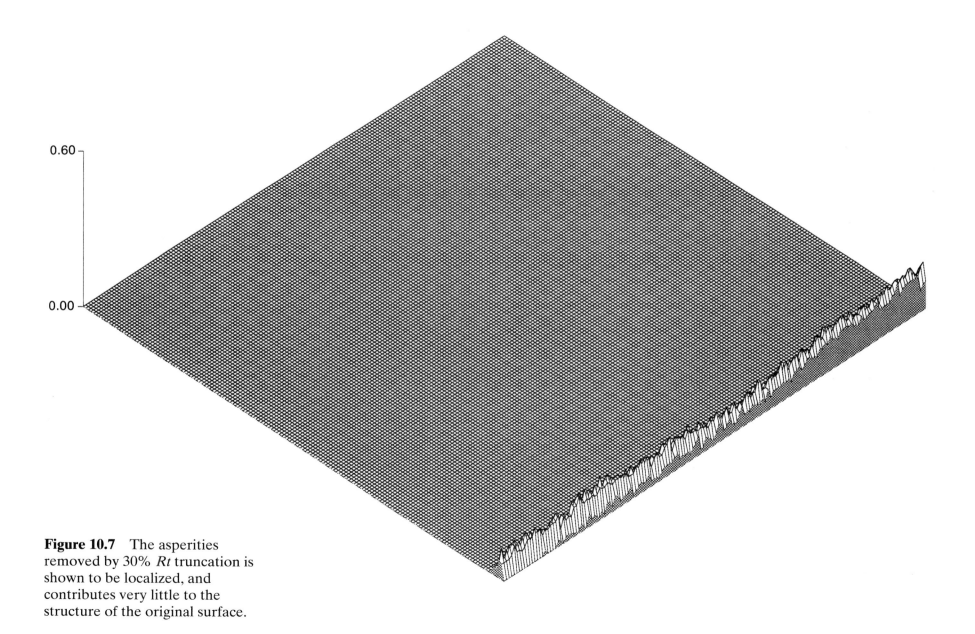

Figure 10.7 The asperities removed by 30% *Rt* truncation is shown to be localized, and contributes very little to the structure of the original surface.

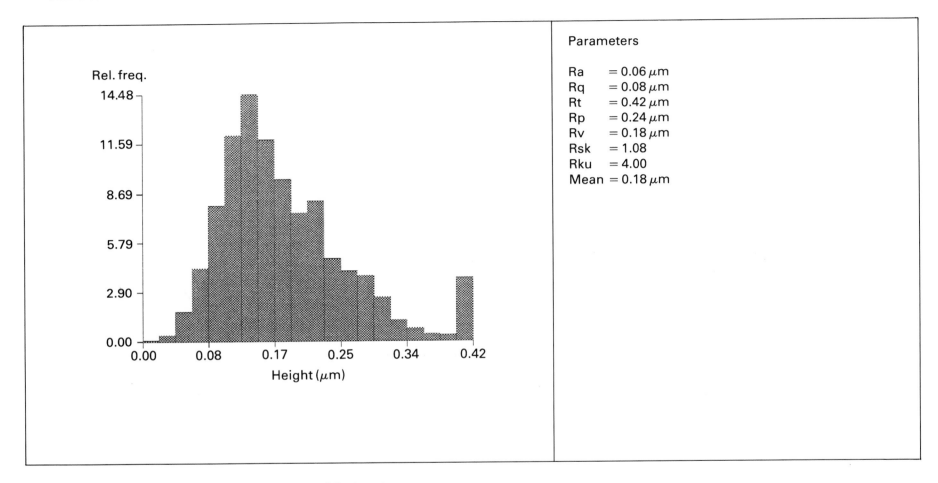

Figure 10.8 The height parameters of the modified surface are affected very little by the process of truncation. Notice that although *Rt* has reduced significantly *Ra*, *Rsk* and *Rku* have changed very little.

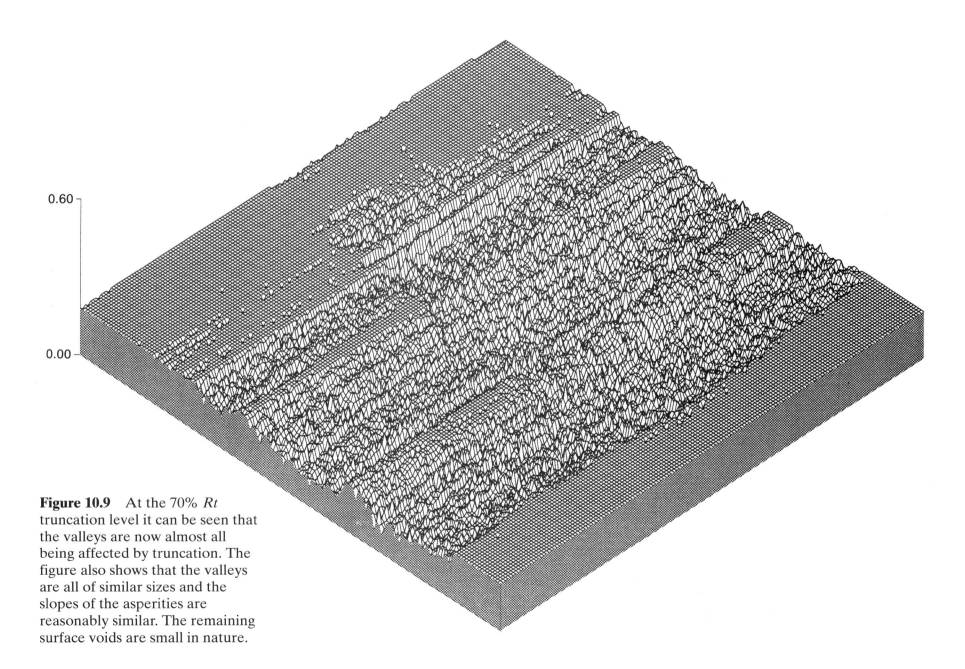

0.60 —

0.00 —

Figure 10.9 At the 70% *Rt* truncation level it can be seen that the valleys are now almost all being affected by truncation. The figure also shows that the valleys are all of similar sizes and the slopes of the asperities are reasonably similar. The remaining surface voids are small in nature.

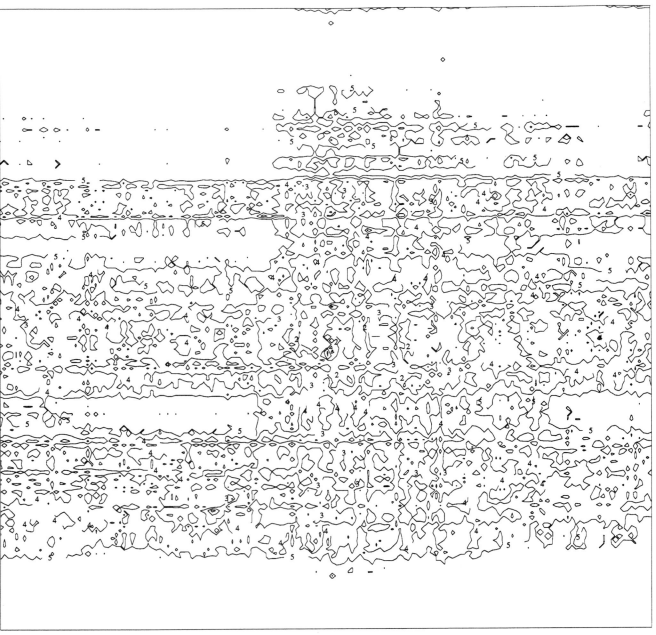

Contour key (μm)

1 : 0.02
2 : 0.05
3 : 0.09
4 : 0.13
5 : 0.16

Figure 10.10 The contour map for the polished surface truncated to 70% *Rt* shows that large flat areas will have the effect of reducing loading pressures when the surface is brought into contact with a second surface.

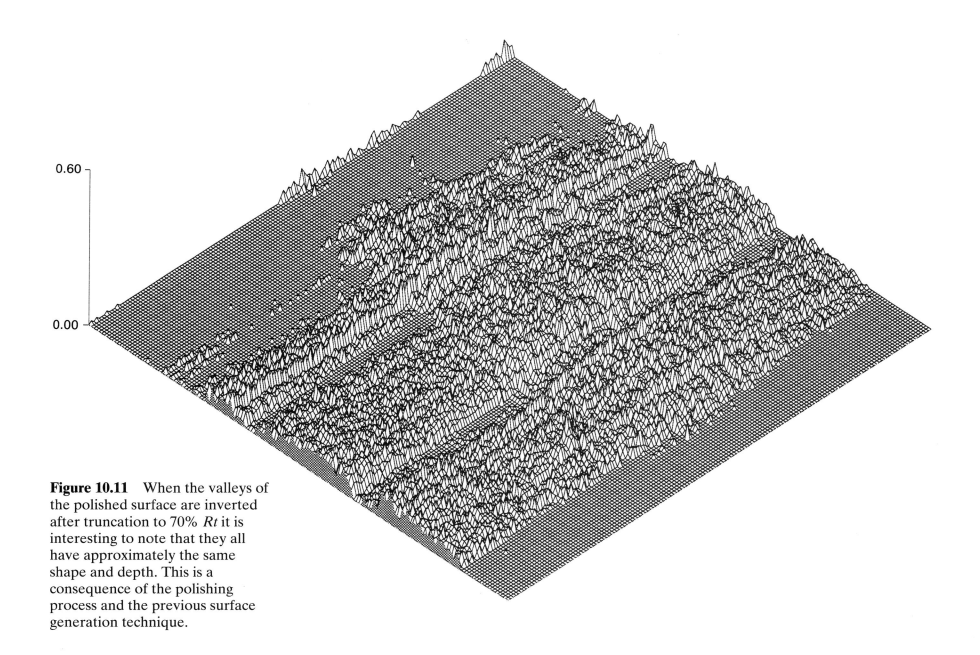

0.60 –

0.00 –

Figure 10.11 When the valleys of the polished surface are inverted after truncation to 70% *Rt* it is interesting to note that they all have approximately the same shape and depth. This is a consequence of the polishing process and the previous surface generation technique.

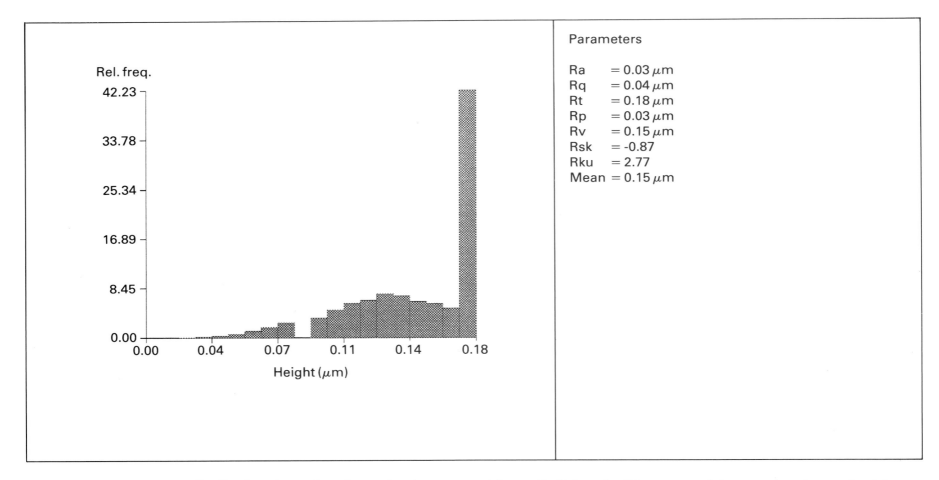

Figure 10.12 The height distribution shows that the surface is becoming significantly skewed at 70% *Rt* truncation; and most of the uppermost asperities have been removed to reveal a flat plane.

Figure 10.13 (over) The nature of the curve on the graph of the plot of contact % against truncation % for the polishing process indicates that the contact % initially changes very little as the surface is truncated. This is because the polishing process affects the shape (form) of the surface, hence the simulated wear removes just the highest asperities on the edges of the surfaces (see Figure 10.1).

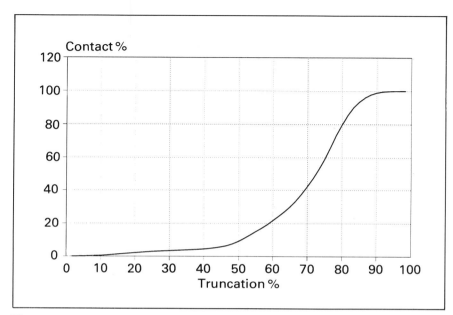

Figure 10.13

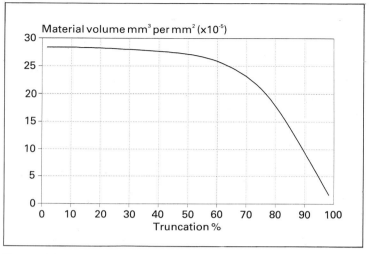

Figure 10.14

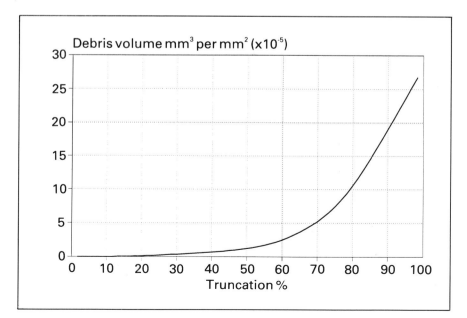

Figure 10.15

Figure 10.14   This graph shows material volume remaining above the lowest asperity valley plotted against truncation level. Since the surface exhibits a significant waviness caused by the action of polishing, truncation causes very little reduction of material volume until the 50% Rt truncation level is reached. After 50% Rt truncation the reduction of volume becomes more significant, but in real terms remains relatively small until the 70% Rt truncation level is reached. After this stage, much more of the surface is being truncated, the effect becoming linear until all the remaining valleys are removed.

Figure 10.15   The debris volume plotted against truncation % shows that at the early stages of truncation, as would be expected, very little debris volume is produced, since the areas in contact successively with the truncation plane are very small. At approximately 60% Rt truncation the debris volume curve starts to increase, with the increase becoming significant at the 70% Rt level. From this level, as has been established in earlier figures, much more of the surface is coming into contact with the truncation plane, the increase being linear at this stage. Since the general roughness of the surface is very small the total wear takes place over a very limited time period in a tribological situation.

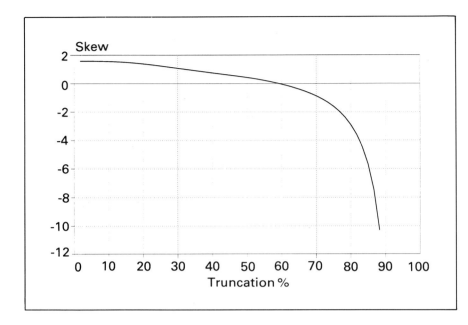

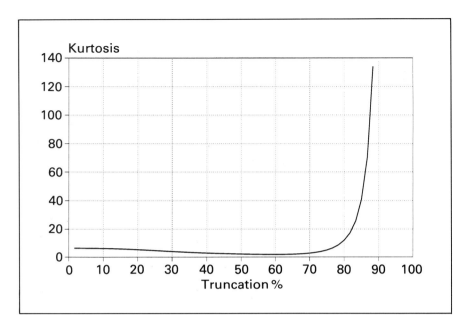

Figure 10.16   This graph indicates that the skewness develops from an initially positive value approaching 2 to a value of zero at 59% *Rt* truncation. The skewness continues to develop an increasingly negative value reaching –2 at 76% *Rt* truncation and rising to –4 at 83%. As with most surfaces the value of skewness rapidly increases from this point.

Figure 10.17   The curve of truncation level against kurtosis shows that the value of kurtosis reduces initially but tends to increase its value at 63% *Rt* truncation. At 76% *Rt* truncation the value of kurtosis starts to increase rapidly, the increase becoming almost linear at the 85% level.

11

CYLINDRICALLY GROUND SURFACES

The figures presented in this section relate to a surface which is typical of those produced by the cylindrical grinding process. The axonometric projection (Figure 11.1) is a section from a cylindrically ground surface which has had the curvature removed by data manipulation to assist visualization. The surface indicates quite clearly the result of the action of the multiple point cutting process. It exhibits three distinct feed marks associated with the feed of the grinding wheel as it passes across the specimen during machining. Superimposed on the feed marks, which are usually described as the waviness within a surface, is the micro-roughness. The micro-roughness is associated with the cutting action that occurs at the interface between the abrasive grits within the grinding wheel and the chip, and is complicated by the variation in heights of the grits within the structure of the grinding wheel. This variation is generally random by nature. These micro-roughness components of the surface are also associated with the non-homogeneity within the body of the material itself which contributes to a variation in the cutting forces and hence the asperity levels within the surface. Some of the spikes which are seen on the surface may be caused by re-welded metal particles or, alternatively, small particles of grinding grit on to the welded machined surface. These phenomena have been established by subjecting ground surfaces to scanning electron microscopy to determine the nature of the adhered particles.

When the valleys are inverted (Figure 11.11) it becomes evident that the majority are of a similar size. Only very few valleys show significant depth characteristics, and those deep valleys which do exist are well distributed across the regions of remaining asperities. This indicates that similar statistical events randomly cause these valleys, possibly these events relate to asperity welding and subsequent metallic tearing from the surface.

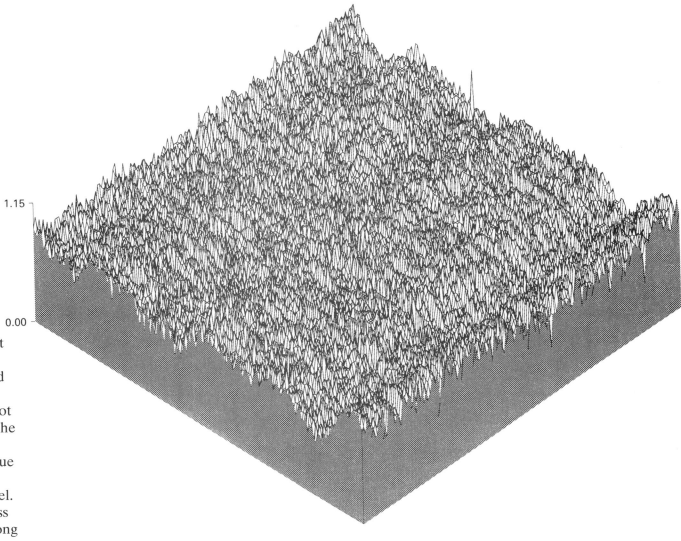

1.15

0.00

Figure 11.1 Axonometric projection with the curvature removed. Note the three distinct feed marks, which are clearly visible despite the superimposed micro-roughness. Although the projection in Figure 11.1 does not indicate much directionality in the cylindrically ground surface, a directionality effect is present due to the action of the diamond which dressed the grinding wheel. As a consequence, the roughness is greater across the lay than along it. The skewness and kurtosis follow the same trend, although to a lesser extent. It should be noted that the process of cylindrical grinding yields almost zero skewness at the 'as machined' condition.

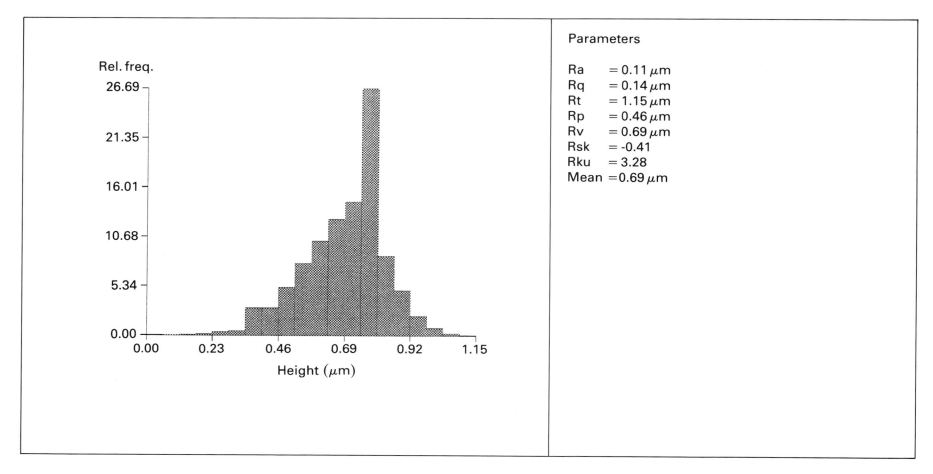

Parameters

Ra   = 0.11 μm
Rq   = 0.14 μm
Rt   = 1.15 μm
Rp   = 0.46 μm
Rv   = 0.69 μm
Rsk = -0.41
Rku = 3.28
Mean = 0.69 μm

Figure 11.2 The histogram represents the amplitude distribution of the asperities within the 3-D surface shown in Figure 11.1. The shape parameters of the surface, skewness and kurtosis, indicate that the cylindrical grinding process produces a pseudo-Gaussian distribution of asperities on the surface (*Rsk* = –0.41; *Rku* =3.28). The difference between these values and the exact values for a Gaussian surface are not significant. The roughness of the surface is small (*Rt* = 1.15 μm; *Ra* = 0.11 μm).

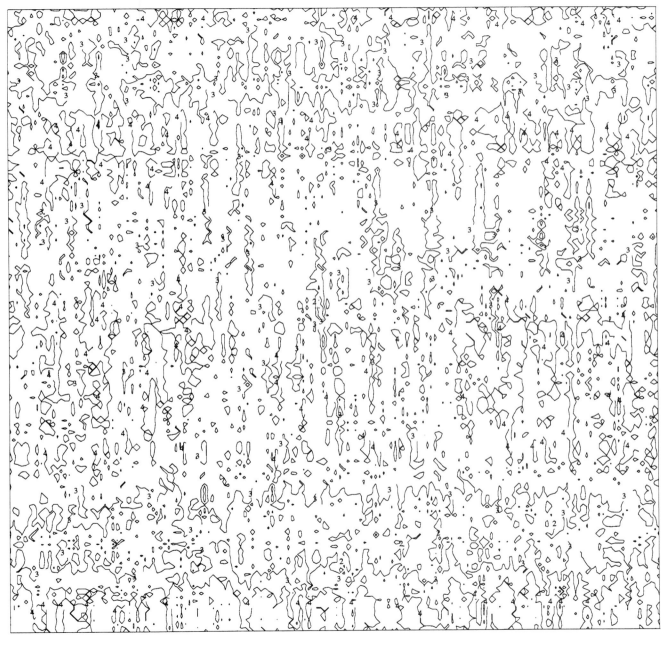

Contour key (μm)

1 : 0.11
2 : 0.34
3 : 0.58
4 : 0.81
5 : 1.03

Figure 11.3 The contour map indicates that although feed marks were observed in Figure 11.1, this feature is not visually obvious in this form of visual presentation. The map shows that the surface is well distributed at all the contour levels shown in the figure.

Figure 11.4 (over) Curves for the parameters Ra', Rq', skewness and kurtosis of 2-D traces taken at orthogonal directions over the area shown in Figure 11.1.

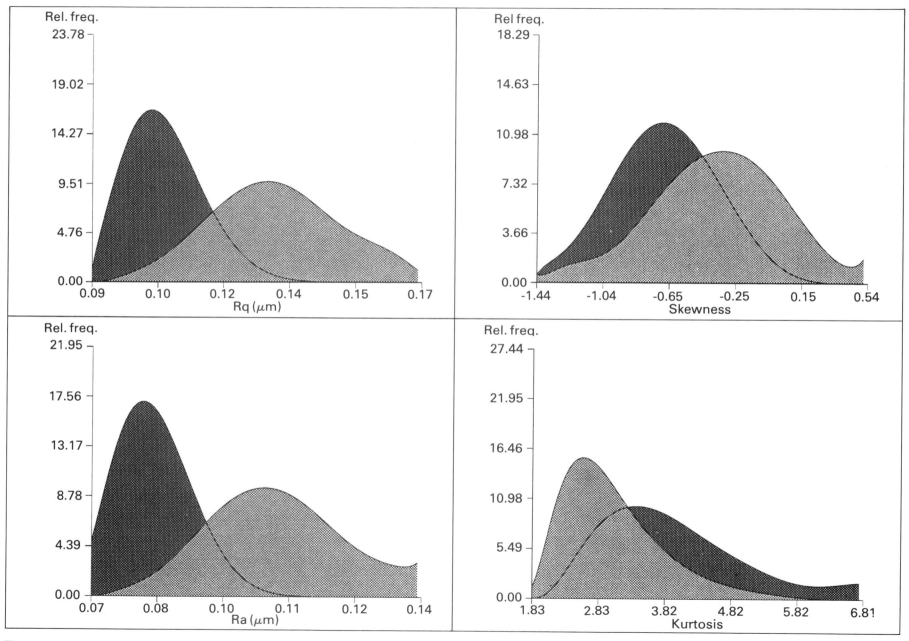

Figure 11.4

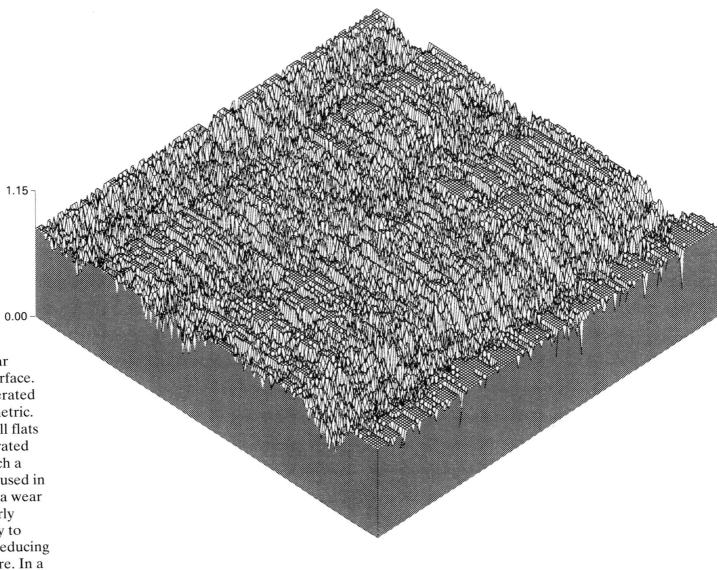

1.15 —

0.00 —

Figure 11.5 At 30% *Rt* truncation small flats appear regularly over the entire surface. This indicates that the generated profile is regular and symmetric. The appearance of the small flats lead to small but well separated contact areas, therefore such a surface may be ideal when used in a wearing environment. In a wear environment significant early plastic deformation is likely to occur having the effect of reducing the asperity contact pressure. In a tribological sense this surface has to wear more than 30% *Rt* before it can be considered 'run-in'. Even so the wear volume would still be very small.

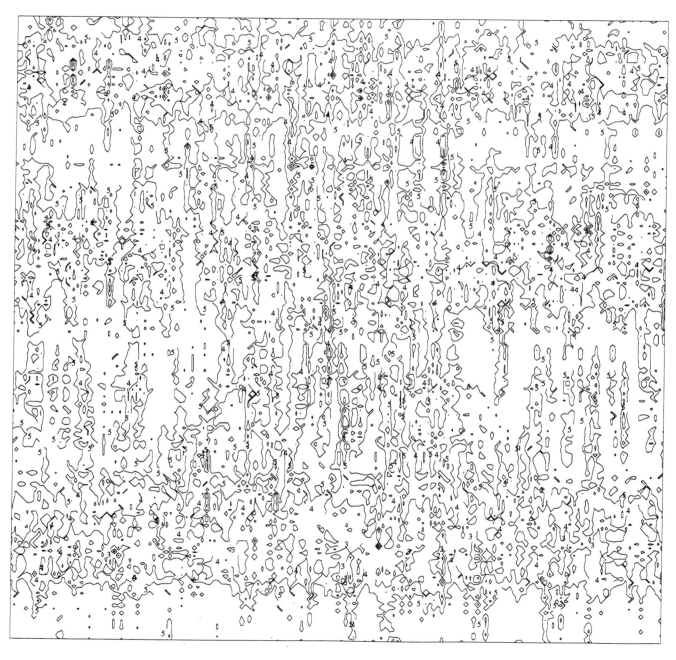

Contour key (μm)

1 : 0.08
2 : 0.24
3 : 0.40
4 : 0.56
5 : 0.72

Figure 11.6 The contour map after the 30% *Rt* truncation is closely similar to the original plot shown in Figure 11.3. This similarity indicates that there is very little structural change within the surface.

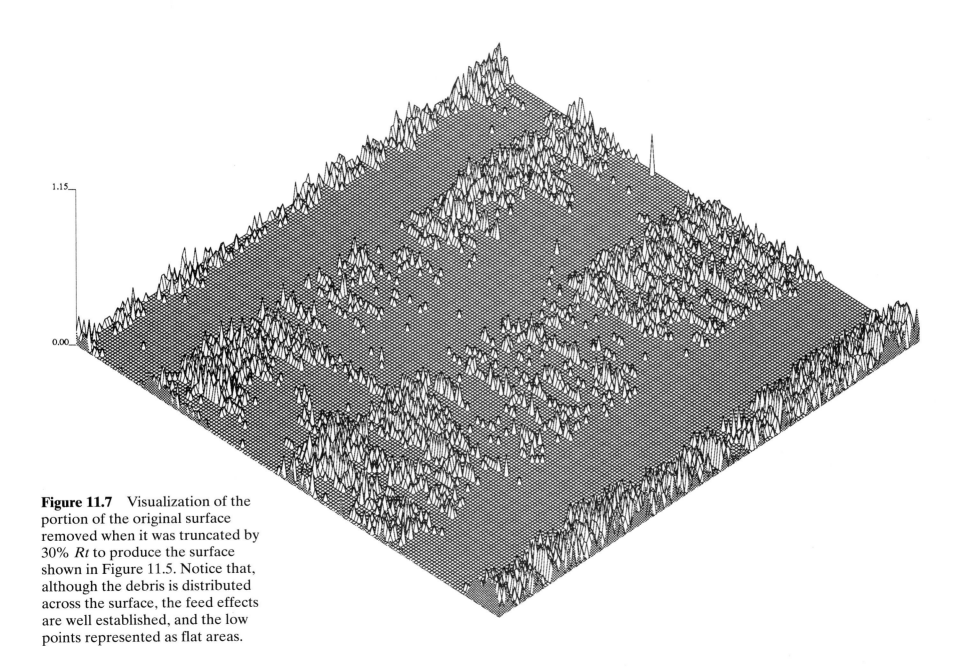

1.15

0.00

Figure 11.7 Visualization of the portion of the original surface removed when it was truncated by 30% *Rt* to produce the surface shown in Figure 11.5. Notice that, although the debris is distributed across the surface, the feed effects are well established, and the low points represented as flat areas.

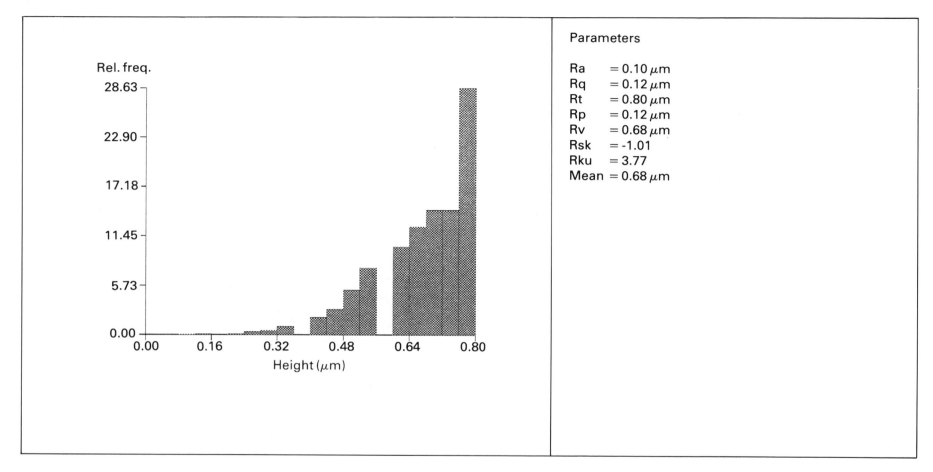

Figure 11.8 The truncation imposed does not significantly affect the surface height parameters. The magnitude of the shape parameters have increased (*Rsk* = –1.01; *Rku* = 3.77).

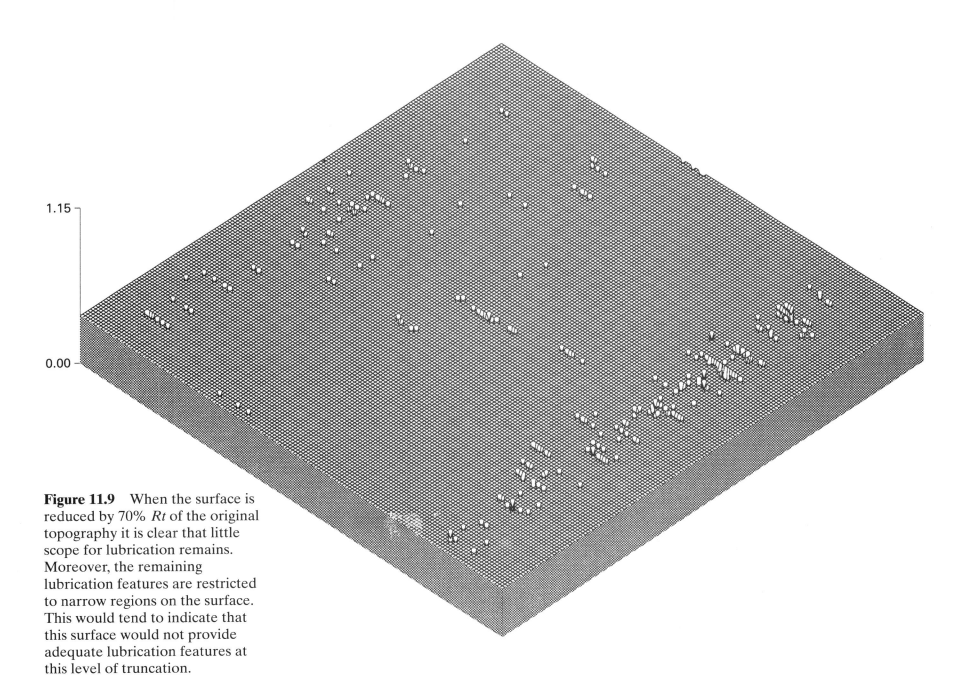

1.15 —

0.00 —

Figure 11.9 When the surface is reduced by 70% *Rt* of the original topography it is clear that little scope for lubrication remains. Moreover, the remaining lubrication features are restricted to narrow regions on the surface. This would tend to indicate that this surface would not provide adequate lubrication features at this level of truncation.

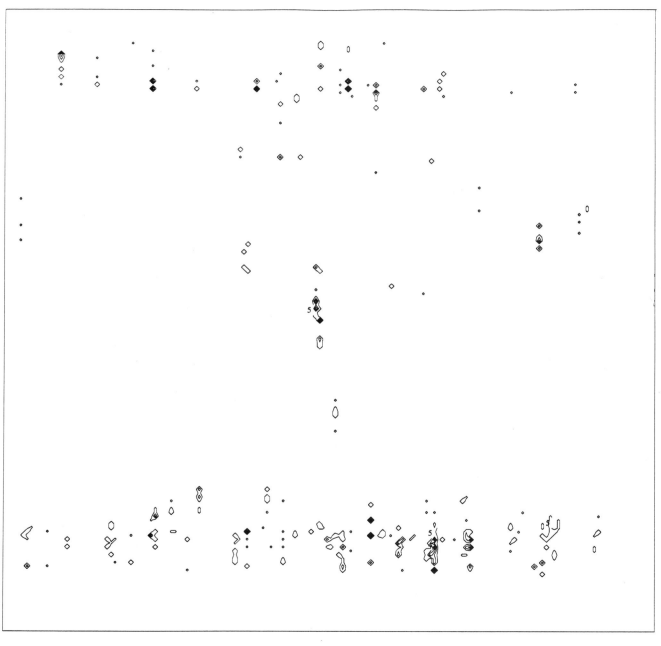

Contour key (μm)

1 : 0.03
2 : 0.10
3 : 0.18
4 : 0.25
5 : 0.31

Figure 11.10 The contour map confirms that at 70% *Rt* truncation, this surface no longer retains adequate lubrication features.

1.15 —

0.00 —

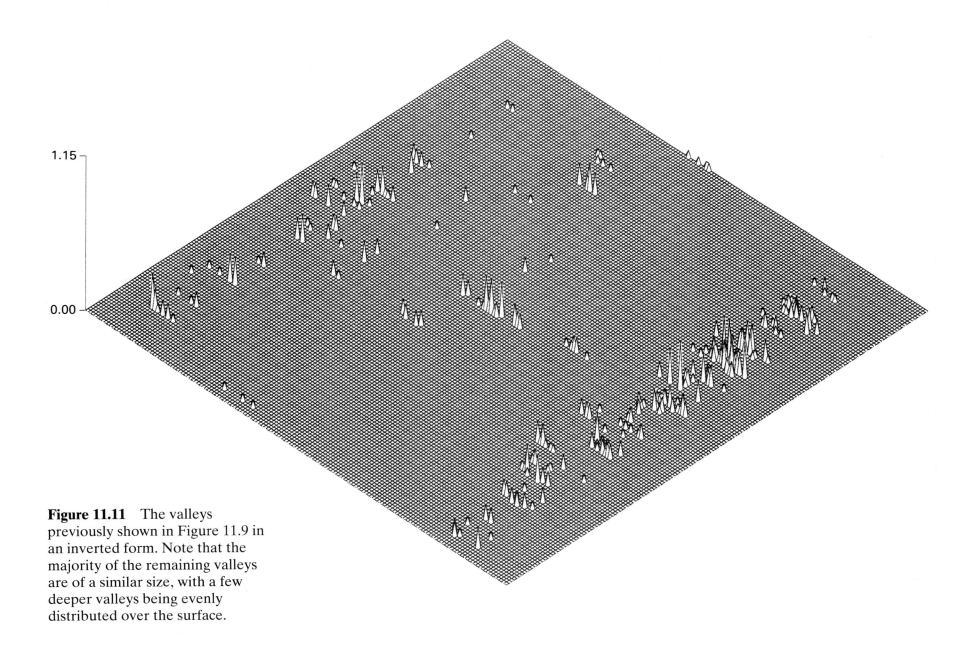

Figure 11.11 The valleys previously shown in Figure 11.9 in an inverted form. Note that the majority of the remaining valleys are of a similar size, with a few deeper valleys being evenly distributed over the surface.

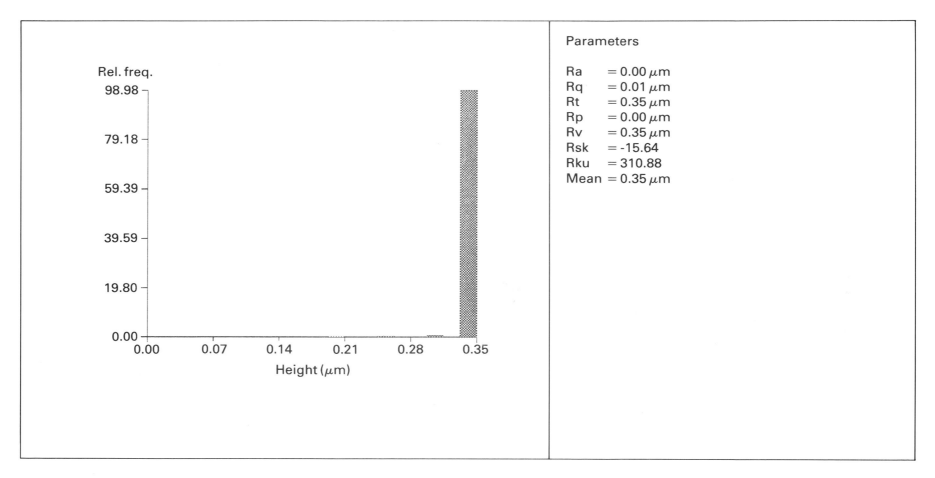

Figure 11.12 When the height distribution of surface asperities of the truncated area are examined it is seen that these cause most of the surfaces to be accumulated at the uppermost level as is evidenced by the impulse on the amplitude distribution curve.

Figure 11.13 (over) A plot of contact % against truncation shows that the increase in contact % remains approximately zero until truncation reaches approximately 15% Rt. At this point the contact % steadily starts to increase reaching a maximum rate of increase at the 25% Rt truncation value. The increase remains approximately constant until the 50% Rt truncation level is reached, at which point the rate of increase diminishes as the valleys become a very small part of the total surface. At 70% Rt truncation the rate of increase of truncation % becomes a minimum, as at this point very few lubricating valleys remain.

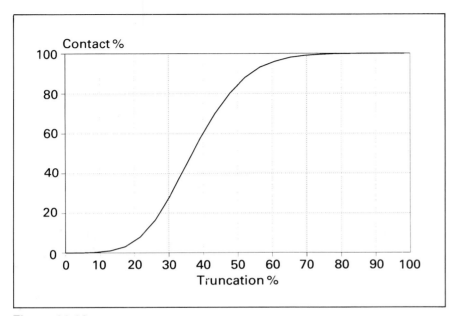

Figure 11.13

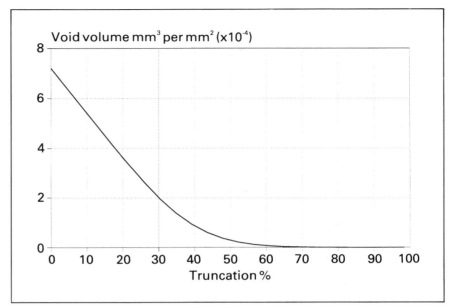

Figure 11.14

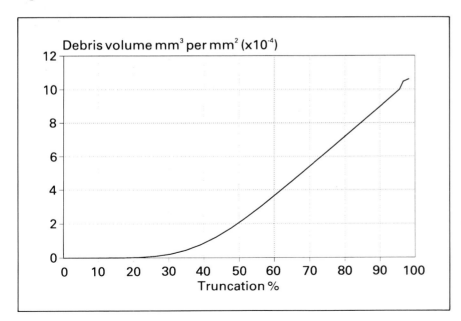

Figure 11.15

Figure 11.14 The void volume parameter shows that the volume of oil decreases rapidly as the uppermost asperities are removed. Since the asperities are very small the oil volume reduction is large and this rate of reduction continues at approximately the same rate until the plane area starts to become large (at approximately the 40% Rt truncation level). At this point the oil volume reduces rapidly until it becomes very small at the 60% Rt truncation level. It is in this region that the surface would begin to be starved of oil retention capacity. Thus, the lubrication retention capacity diminishes rapidly after this point.

Figure 11.15 This graph indicates that no significant debris is produced until the surface is truncated by between 25 to 30% Rt. At this level of truncation, the debris volume from the surface increases until the rate of removal is approximately constant with increasing depth after the 40% Rt truncation level.

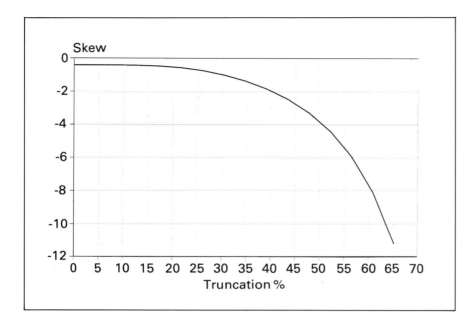

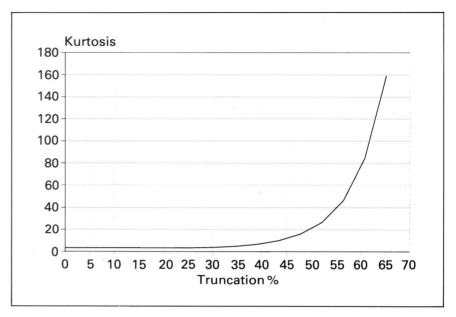

Figure 11.16 This figure shows the surface asymmetry parameter, skewness, progressing with truncation. Initially, the surface exhibits almost zero skewness and as truncation progresses the change in skewness is negligible until the 30% *Rt* truncation level is reached. From this point the rate of increase (negatively) of skewness accelerates until it becomes excessive at 60% *Rt* truncation. At this point it can be expected that the surface will become unsuited to a tribological environment.

Figure 11.17 The curve of kurtosis against truncation shows that in this case, the initial value of kurtosis is very low and does not change substantially until approximately 35% *Rt* truncation. From approximately 55% *Rt* truncation, the value of kurtosis rapidly increases.

12 LAPPED SURFACES

The figures presented in this section relate to a surface which is typical of those produced by the lapping process. Such a surface is generally random in nature. Two other features are also detectable on the surface. These are small flats which have been generated by the action of the lap during the machining process, a mechanism adequately described as controlled and deliberate wearing. Finally a number of individual underlying valleys are produced, approximately equally spaced, although not all of the same depth, which were originally caused by the action of the previous machining process.

The area histogram, presented in Figure 12.2, shows that the lapping process produces a normally distributed surface (see Figure 12.2). Such a condition is ideal for a surface which is to be employed in a tribological situation; a condition under which such surfaces are often used.

The contour map (Figure 12.3) shows that the nature of the cutting process is not directional, since flats are being generated over all sections of the surface. Although this is the case the underlying topography associated with the previous machining process is still in evidence at the lower surface levels. Indeed, once the surface is truncated (Figures 12.5 and 12.6) it becomes clear that the lapping process produces a form error in the surface due to the manual nature of the process. This problem does not exist in well controlled machine lapping. Errors in form, produced as a result of lapping, have been recognized as a serious problem in the production of high-precision ball and roller bearings. It is also of value to reinforce the point that although the form error is included in some of the earlier figures, was obscured from view. An advantage of the truncation process is that it assists in revealing such sub-surface effects.

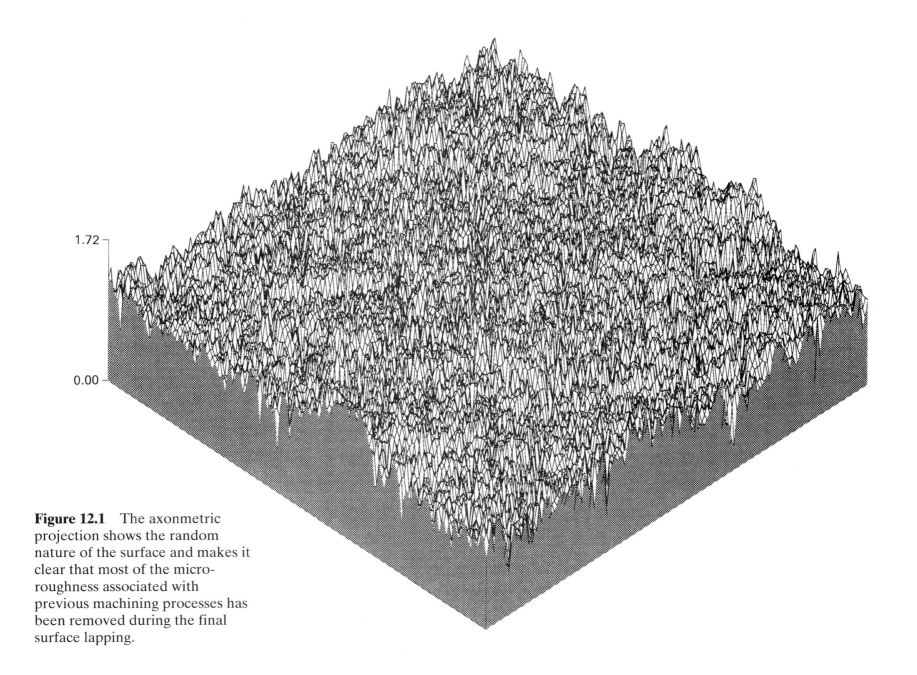

1.72 –

0.00 –

Figure 12.1 The axonmetric projection shows the random nature of the surface and makes it clear that most of the micro-roughness associated with previous machining processes has been removed during the final surface lapping.

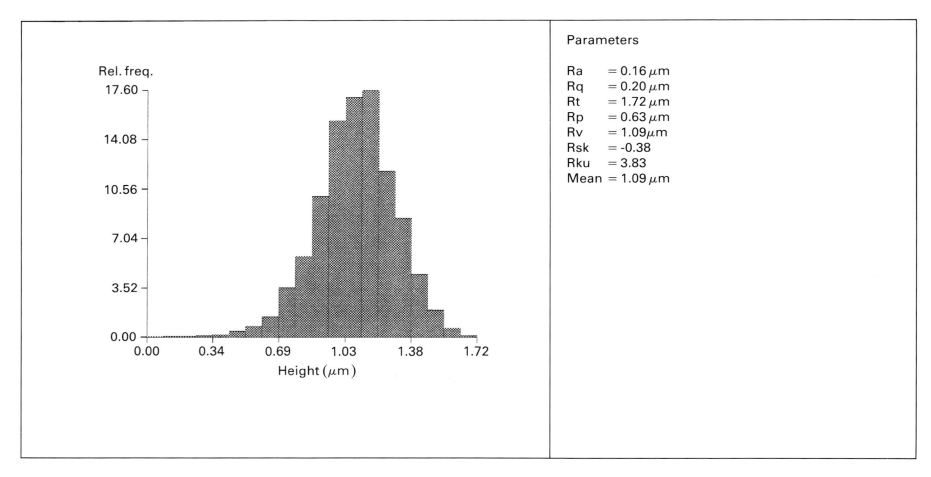

Figure 12.2 The histogram shows that the surface is normally distributed, with skewness, $Rsk = -0.38$. The corresponding kurtosis value is typical, $Rku = 3.83$. The average roughness of the surface has been reduced by the lapping process to a value of $Ra = 0.16$ μm.

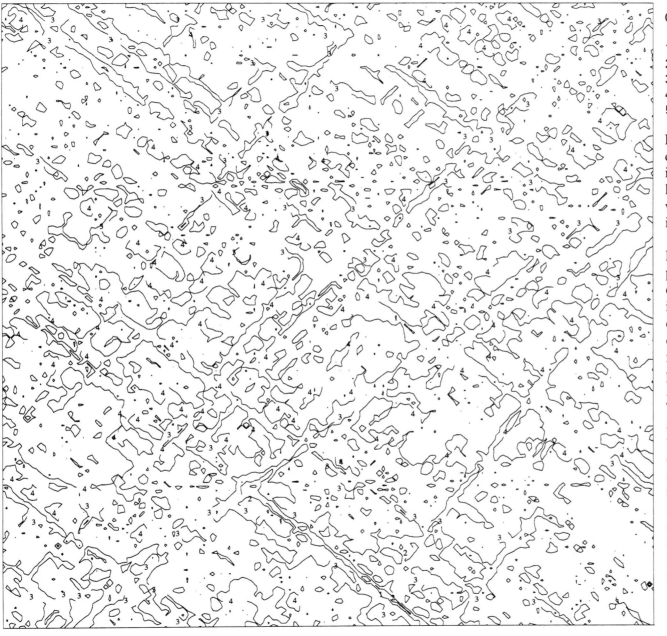

Contour key (μm)

1 : 0.17
2 : 0.52
3 : 0.86
4 : 1.20
5 : 1.55

Figure 12.3 The contour map confirms that the lapping process is not directional. At this initial level the topography associated with previous machining processes is not obvious.

Figure 12.4 (over) This figure presents information on 2-D data extracted from the surface shown in Figure 12.1. Traces have been extracted in two orthogonal directions from the surface. The frequency distribution curves presented relate to the parameters Ra', Rq' skewness and kurtosis. The distribution curves for the asperity height distribution show that there are relatively minor differences between the 'with lay' and 'across lay' parameters. The shape parameters skewness and kurtosis also exhibit little differences between the two directions, the small differences could be explained purely by sampling variations.

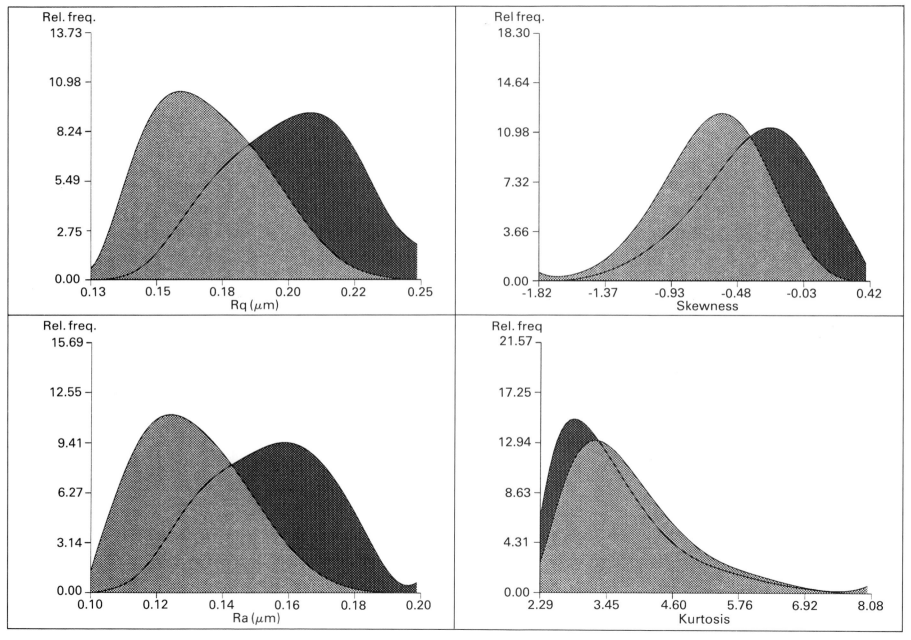

Figure 12.4

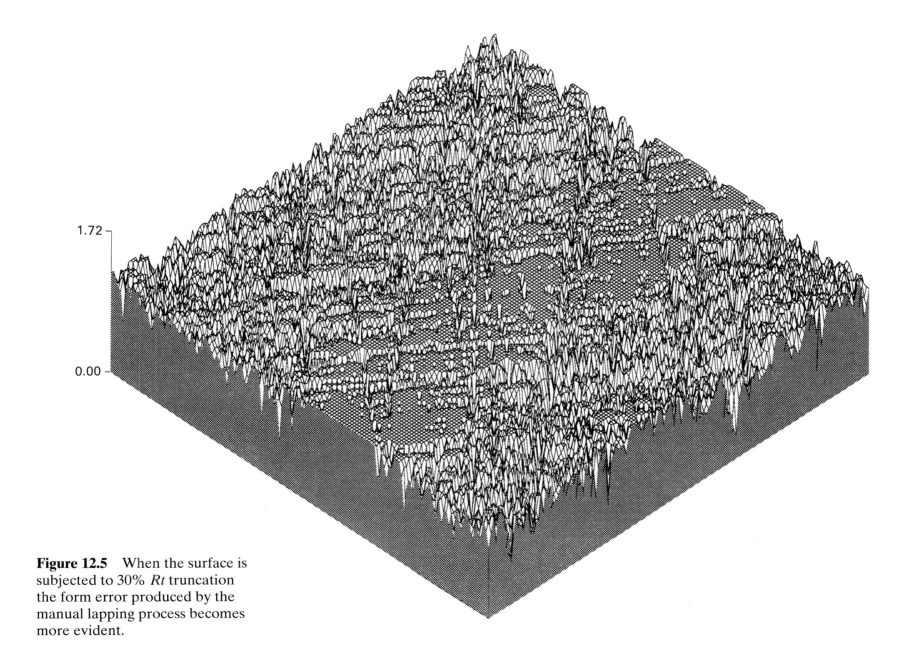

1.72

0.00

Figure 12.5 When the surface is subjected to 30% *Rt* truncation the form error produced by the manual lapping process becomes more evident.

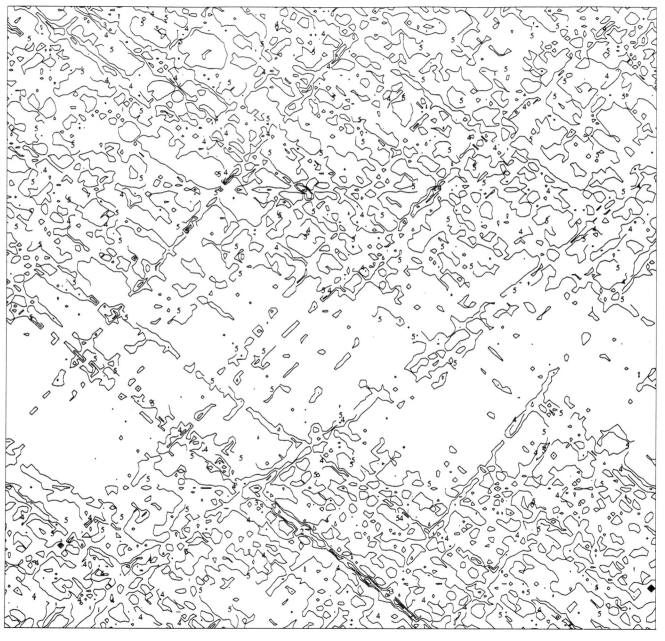

Contour key (μm)

1 : 0.12
2 : 0.36
3 : 0.60
4 : 0.84
5 : 1.08

Figure 12.6 Contour map of the truncated surfaces shown in Figure 12.5: note that the underlying topography which was difficult to detect in Figure 12.3 is now becoming clearer.

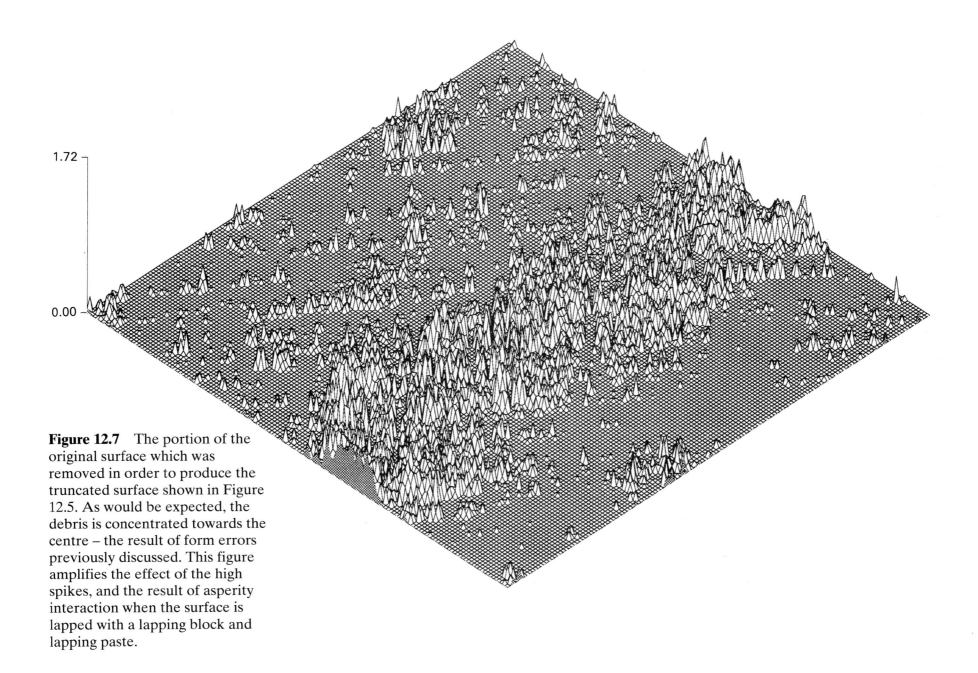

1.72

0.00

Figure 12.7 The portion of the original surface which was removed in order to produce the truncated surface shown in Figure 12.5. As would be expected, the debris is concentrated towards the centre – the result of form errors previously discussed. This figure amplifies the effect of the high spikes, and the result of asperity interaction when the surface is lapped with a lapping block and lapping paste.

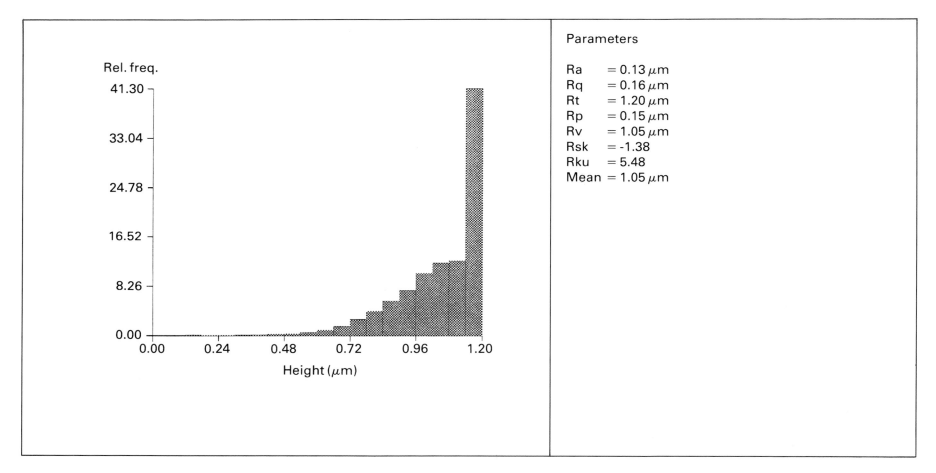

Figure 12.8 The modified 3-D height distribution for the 30% *Rt* truncated surface shows that many of the asperities are concentrated at the uppermost levels. Note that the skewness (*Rsk* = –1.38) has not developed significantly, and its magnitude reflects closely the previously asymmetric value.

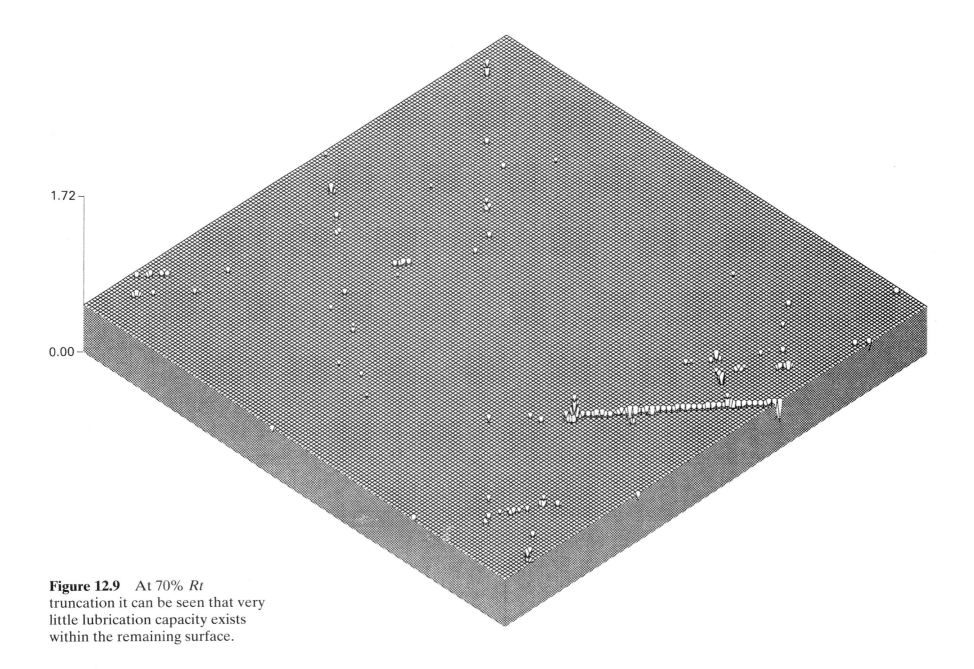

1.72 —

0.00 —

Figure 12.9 At 70% *Rt*
truncation it can be seen that very
little lubrication capacity exists
within the remaining surface.

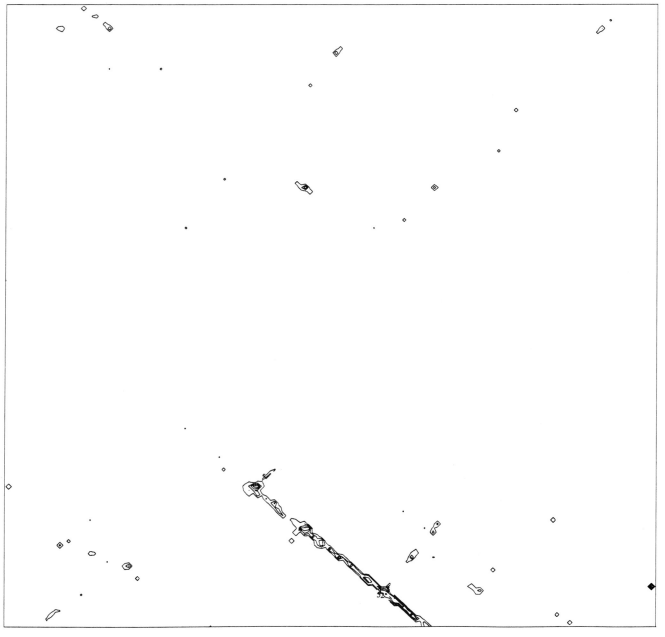

Contour key (μm)

1 : 0.05
2 : 0.16
3 : 0.26
4 : 0.36
5 : 0.47

Figure 12.10 The contour map of the 70% *Rt* truncated surface confirms that the surface is now becoming a near featureless plane.

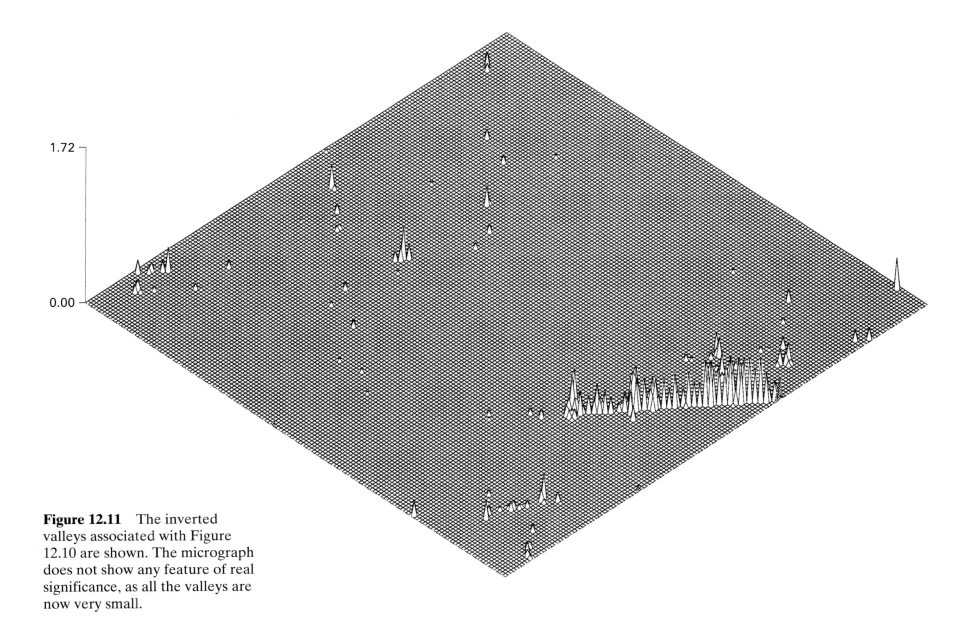

1.72

0.00

Figure 12.11 The inverted valleys associated with Figure 12.10 are shown. The micrograph does not show any feature of real significance, as all the valleys are now very small.

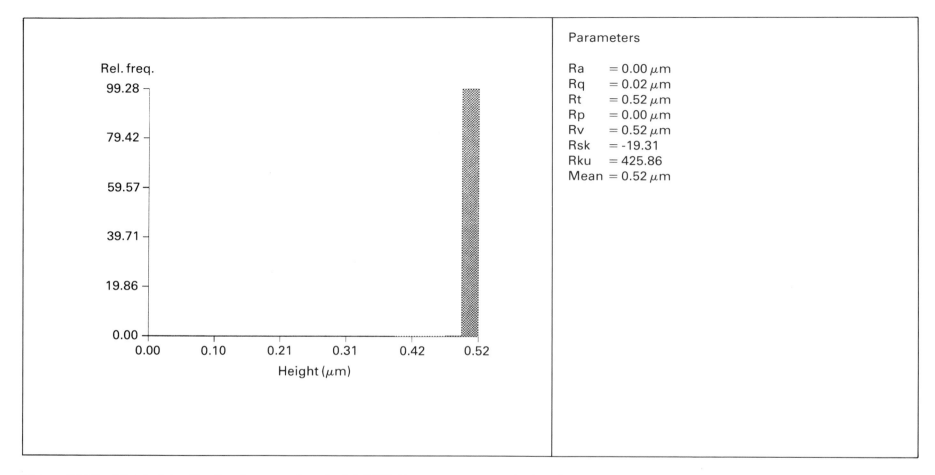

Figure 12.12 The height distribution associated with 70% Rt truncation is dominated by the flat plane and hence the parameters of surface shape, namely skewness and kurtosis, do not provide at this stage any meaningful information which can be used in a functional sense.

Figure 12.13 (over) When considering this curve and those in the following four figures it should be noted that the general scale of roughness is very small and hence the changes which occur do so over a very limited range of truncation or theoretical wear. This graph showing contact % against truncation level demonstrates that wear of the surface rapidly leads to an increase in contact %. Note that at 10% Rt truncation, the contact % is approximately 2%, but by the time truncation has reached 20% Rt the contact % has increased to 8%. From this point the ratio of contact % to truncation % reaches a maximum, and this maximum continues until the truncation % is reached. From this position the body of material in contact with the truncation plane is such that the rate of increase diminishes. Note that at the 60% Rt truncation level virtually all the surface is at the uppermost level.

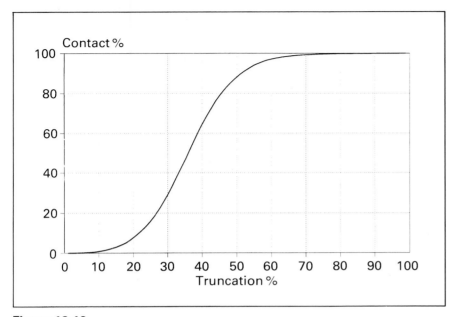

Figure 12.13

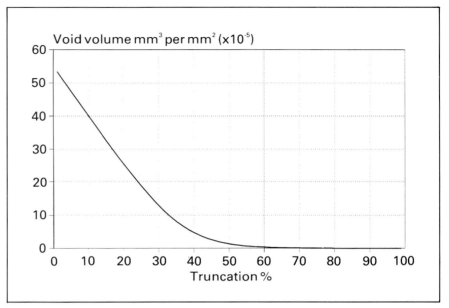

Figure 12.14

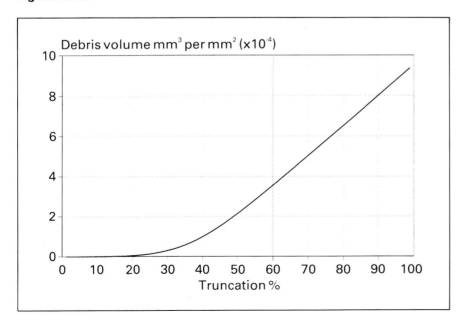

Figure 12.15

Figure 12.14   The curve for void volume against truncation %. This volume is calculated from the highest asperity to the lowest valley on the surface under assessment. At the initial truncation levels the truncation plane removes mostly oil volume as very few asperities interact with the plane. At 40% *Rt* truncation sufficient asperities are now interacting with the plane to minimize the rate at which the oil volume is reducing. When the 60% *Rt* truncation level is reached the oil volume has diminished to the point at which there is insufficient volume to maintain lubrication.

Figure 12.15   From the progression of debris volume plotted against truncation %, it can be seen that from the 35% *Rt* truncation level the rate of increase in debris is constant. Due to the scale of the micro-roughness, the volume associated with debris is small overall for lapped surfaces.

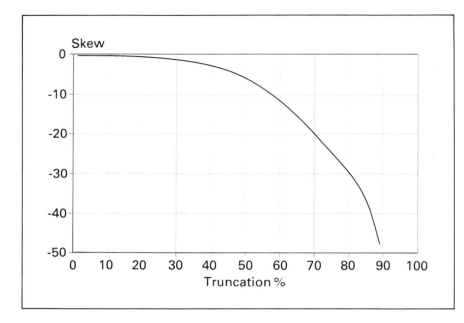

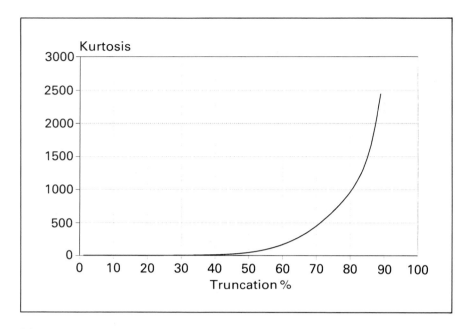

Figure 12.16   The first of the shape parameters, skewness is presented: this parameter starts from an initial value of Rsk = –0.38, and this value remains relatively constant until the 30% Rt truncation level is reached. At this point the skewness parameter starts to increase rapidly (in a negative sense) and it becomes too large to yield a tribological surface at the 60% Rt truncation level.

Figure 12.17   This figure presents information on the second shape parameter, kurtosis. Typically, this parameter is insensitive in the normal operating range of the surface. The value of the parameter at its initial condition, Rku = 3.83, remains approximately constant until the 40% Rt truncation level is reached. The value of kurtosis becomes meaningless after the 60% Rt truncation level.

13 HONED SURFACES

The figures presented in this section relate to a surface which is typical of those produced by the honing process. The surface has a random structure, with superimposed honing grooves running across it. These grooves are produced by the action of specially sharpened honing stones which are traversed and reciprocated under controlled action within a bore as it goes through its final machining process. These honing grooves are introduced to deliberately modify the surface by providing well spaced lubricant retaining features. As the honing stock traverses through the bore the honing grooves become quite deep, but the grooves, which are machined on the return stroke, are considerably less deep; a well known feature of the honing process. The contour map for this honed surface (Figure 13.3) shows clearly the directional effects associated with the honing process, and this method of visual presentation is well suited to establishing the 'lay' of the surface.

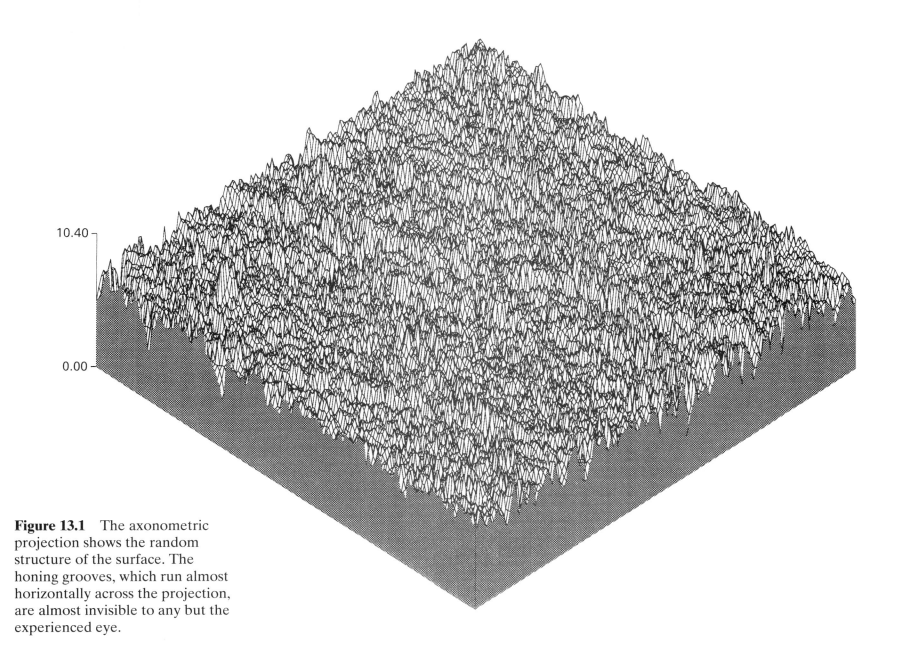

10.40 –

0.00 –

Figure 13.1 The axonometric projection shows the random structure of the surface. The honing grooves, which run almost horizontally across the projection, are almost invisible to any but the experienced eye.

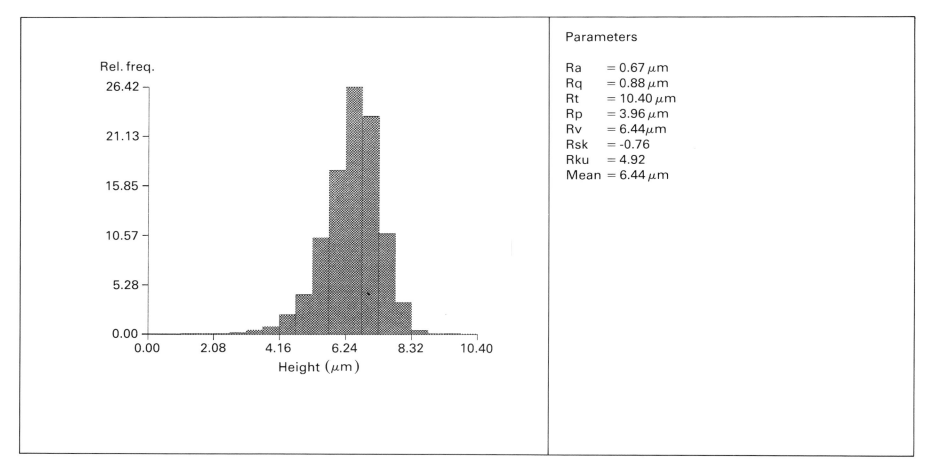

Figure 13.2 The height parameters of the 3-D surface indicate that a honed surface is slightly skewed (*Rsk* = –0.76; *Rku* = 4.92).

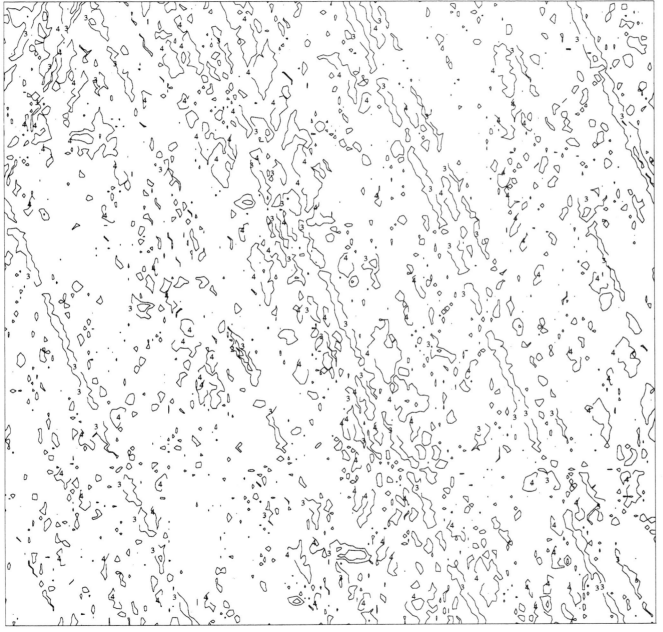

Contour key (μm)

1 : 1.04
2 : 3.12
3 : 5.20
4 : 7.28
5 : 9.36

Figure 13.3 The directional effects of the honing process, which were difficult to detect in Figure 13.1 are much more clearly visible on the contour map.

Figure 13.4 (over) Information related to the 2-D data extracted in the two orthogonal directions. The composite graphs indicate that there are no directional properties within the surface. However, reference to Figure 13.3 shows that directional effects do exist at approximately 30° to the direction of measurement. This indicates a short-coming when investigating directional features using the approach adopted in this figure. When using this technique, measurement should be made either in line with, or normal to, the direction of machining.

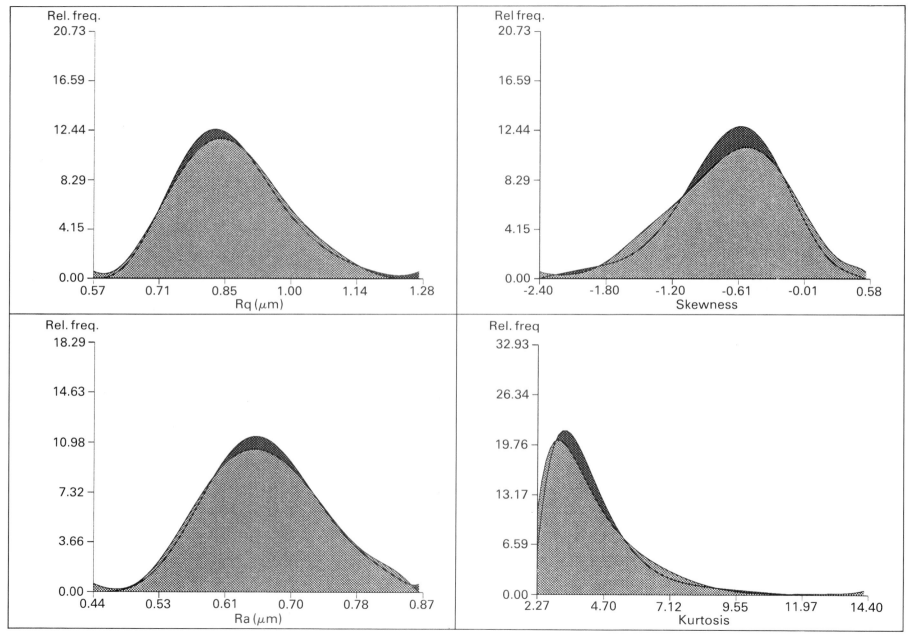

Figure 13.4

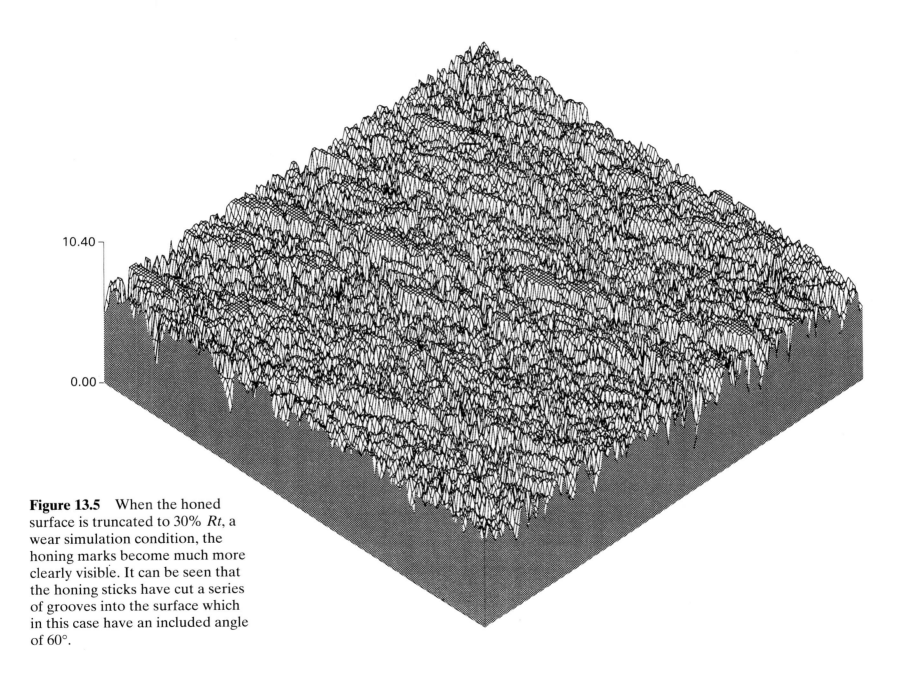

10.40 —

0.00 —

Figure 13.5 When the honed surface is truncated to 30% *Rt*, a wear simulation condition, the honing marks become much more clearly visible. It can be seen that the honing sticks have cut a series of grooves into the surface which in this case have an included angle of 60°.

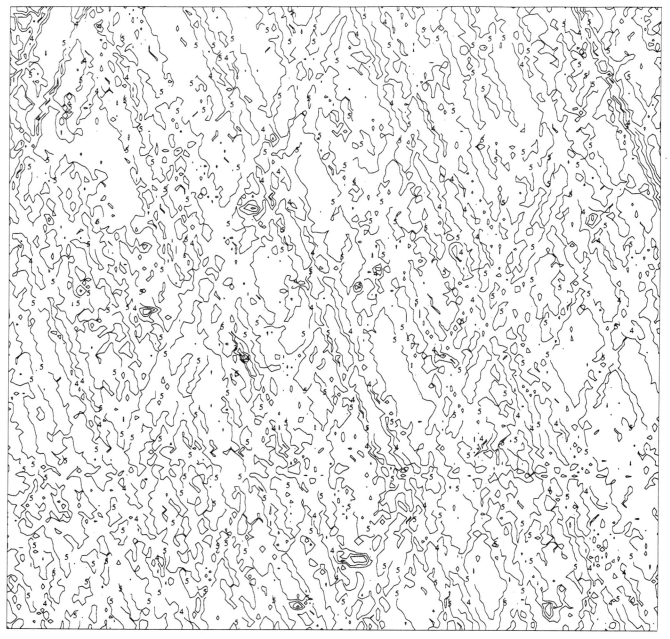

Contour key (μm)

1 : 0.73
2 : 2.18
3 : 3.64
4 : 5.09
5 : 6.55

Figure 13.6 In this contour map of the 30% *Rt* truncated surface, the honing marks are now clearly visible.

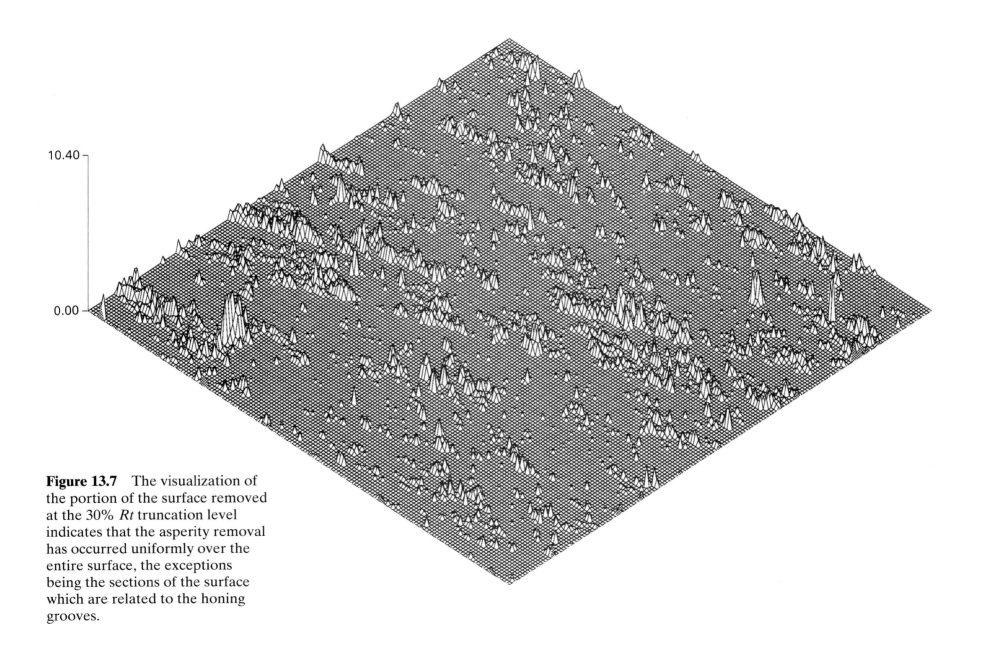

10.40 —

0.00 —

Figure 13.7 The visualization of the portion of the surface removed at the 30% *Rt* truncation level indicates that the asperity removal has occurred uniformly over the entire surface, the exceptions being the sections of the surface which are related to the honing grooves.

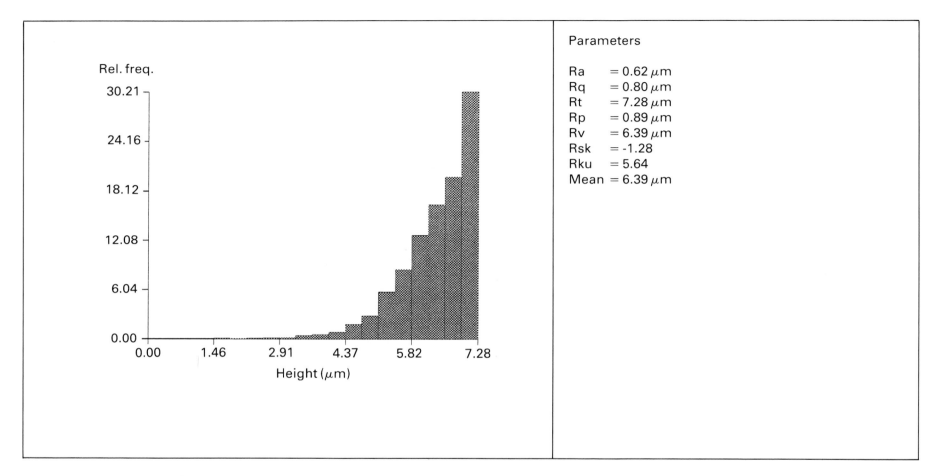

Figure 13.8 The area height distribution parameters associated with the 30% *Rt* truncation condition show that the values are severely affected by this level of truncation, and as a consequence much of the surface is concentrated in the uppermost levels. Note than an impulse is beginning to form at the right hand side of the distribution: a feature which is caused by the proportion of asperities at the uppermost levels of the surface. As a consequence, the skewness of the surface has increased to *Rsk* = –1.28 from its initial value, *Rsk* = –0.76, and the kurtosis is increased from *Rku* = 4.92 to *Rku* = 5.64.

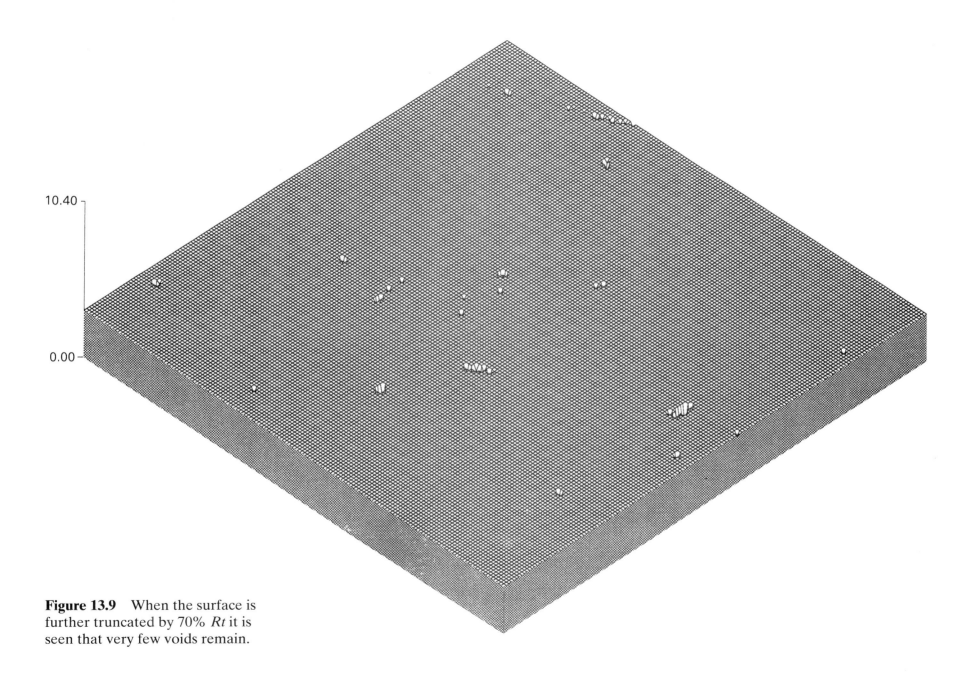

10.40 —

0.00 —

Figure 13.9 When the surface is further truncated by 70% *Rt* it is seen that very few voids remain.

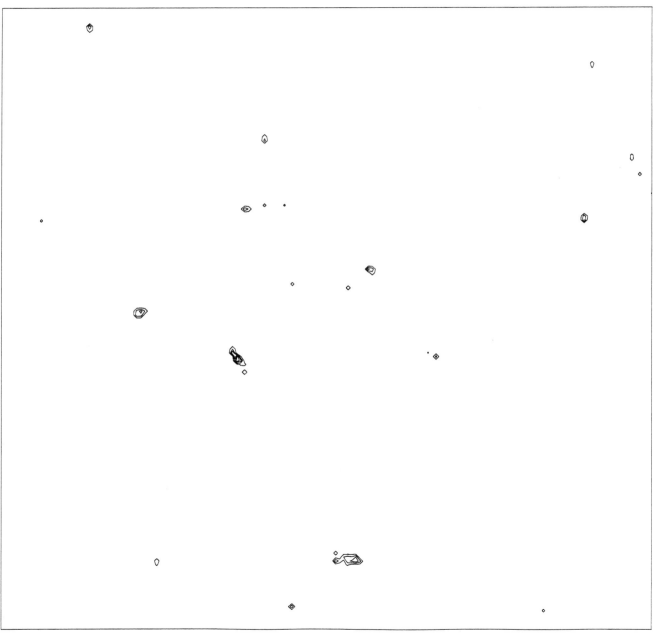

Contour key (μm)

1 : 0.31
2 : 0.94
3 : 1.56
4 : 2.19
5 : 2.81

Figure 13.10 The contour map of the surface after 70% *Rt* truncation shows that the voids which do exist are well distributed. They are, however, probably insufficient at this level to enable adequate lubrication to occur if the surface is used in a tribological environment.

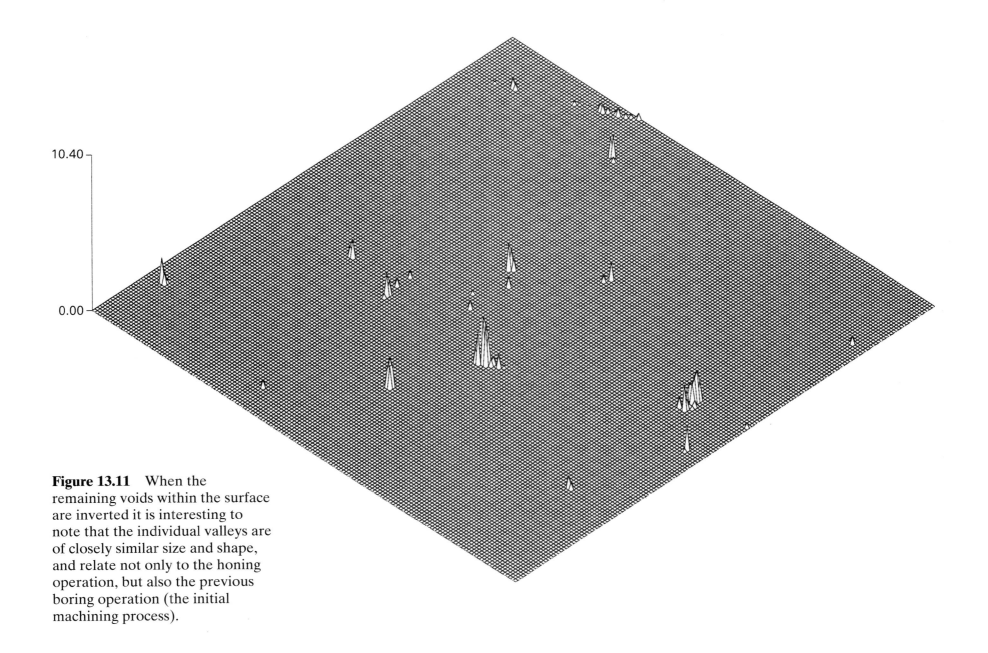

10.40

0.00

Figure 13.11 When the remaining voids within the surface are inverted it is interesting to note that the individual valleys are of closely similar size and shape, and relate not only to the honing operation, but also the previous boring operation (the initial machining process).

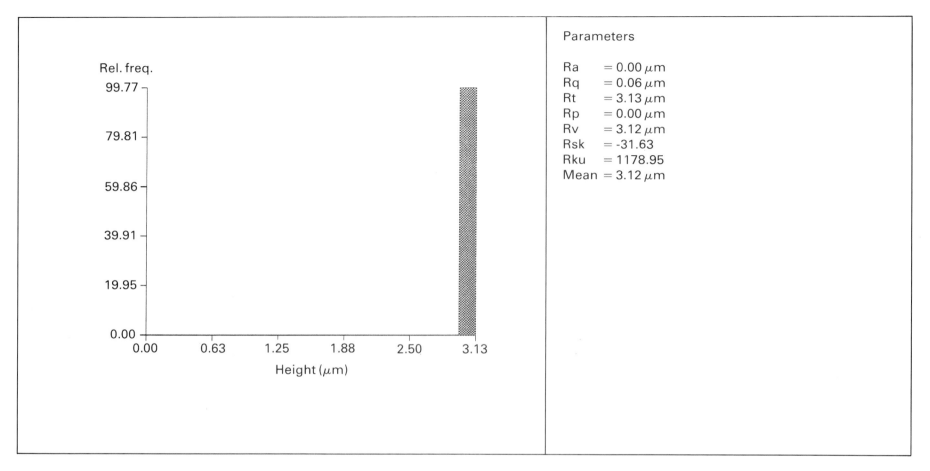

Parameters

Ra   = 0.00 μm
Rq   = 0.06 μm
Rt   = 3.13 μm
Rp   = 0.00 μm
Rv   = 3.12 μm
Rsk = -31.63
Rku = 1178.95
Mean = 3.12 μm

Figure 13.12 The asperity height distribution for the surface truncated by 70% *Rt*. Notice that at this truncation level almost the entire surface is at the uppermost level and this leads to an extreme skewness and kurtosis (*Rsk* = –31.63; *Rku* = 1178.95) which implies that the surface can no longer be regarded as tribologically suitable.

Figure 13.13 (over) A graph of truncation % plotted against contact % shows that initially, as the uppermost asperities are removed, the contact % is almost non-existent. This situation continues until approximately the 22% *Rt* truncation level is attained. As the surface is further truncated the contact % rapidly increases until it reaches the maximum rate, approximately 30% *Rt*. The rate of the increase of contact % starts to fall away at 50% *Rt* truncation, becoming virtually a constant level at the 60% value. From this point the surface cannot retain a tribological function.

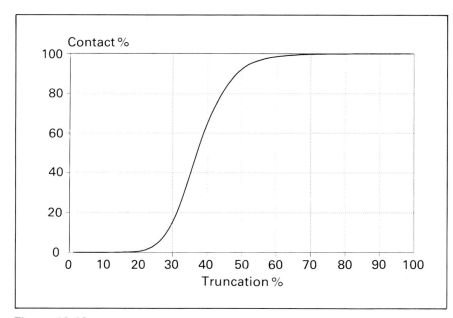

Figure 13.13

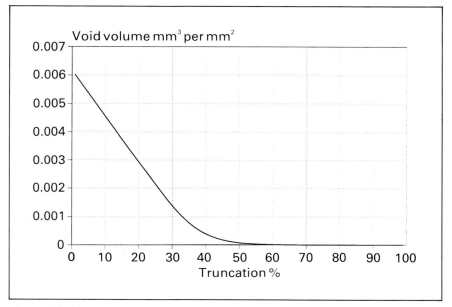

Figure 13.14

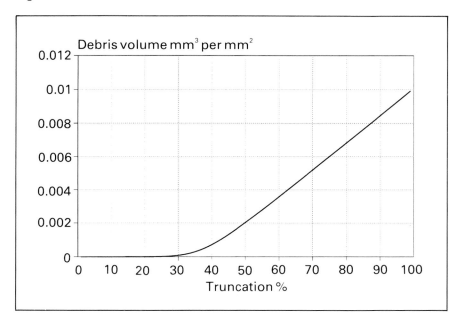

Figure 13.15

Figure 13.14   Truncation % against the void volume parameter. The void volume is specified as the volume beneath the highest asperity to the lowest valley. As the asperities are truncated the void volume reduces rapidly and the rate of decrease remains constant until a fall off starts to show at the 35% *Rt* truncation level. After this level, the decrease in void volume falls off rapidly until it becomes almost non-existent at the 60% *Rt* truncation level.

Figure 13.15   As would be expected with a plot of debris volume versus truncation, the debris volume is insignificant at the lower stages of truncation. At approximately 35% *Rt* truncation the volume of debris starts to become significant, reaching a maximum rate of generation at about 42% *Rt*. The rate of generation continues at the same rate until all valleys are removed.

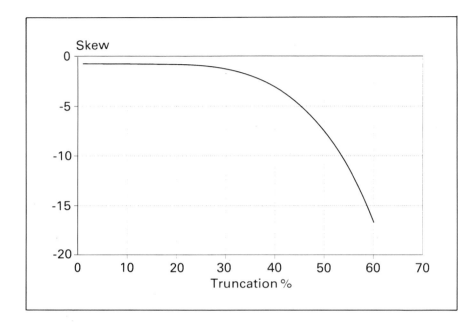

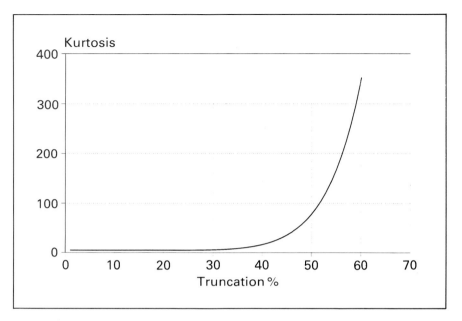

Figure 13.16 This figure illustrates the progression of the shape parameter, skewness. The initial value of skewness remains substantially unaltered until the 35% *Rt* truncation level is reached. From this point on the value steadily increases in a negative sense until it reaches *Rsk* = –17 at 60% *Rt* truncation. The skewness value of *Rsk* = –17 indicates that the surface is no longer suited to a tribological application.

Figure 13.17 The progression of the second surface shape parameter, kurtosis: it has an initial value of *Rku* = 4.92 and this value remains substantially unchanged until a truncation level of 40% *Rt* is reached. The kurtosis value rapidly increases past this point and becomes practically meaningless (supporting the view formed in earlier figures) by the time 60% *Rt* truncation is reached as the surface fails to have any tribological significance.

14 PLATEAU HONED SURFACES

The figures presented in this section relate to a surface which is typical of those produced by the plateau honing process. The plateau honing process is an ensemble of three complementary processes, and its overall character is a combination of appropriate proportions of those three processes. The three processes involved in producing the final surface are boring, followed by honing and then finishing by plateau honing.

The honing process, as demonstrated in Section 13, introduces the honing grooves, which are provided primarily as lubricant retaining features. The machining conditions set during the honing stage of the surface generation determines the angle of the honing grooves. There is much disparity of opinion as to the best angle for these grooves, but in practice they are usually set between 30° and 60° to the vertical.

The plateau honing process is achieved by the use of very finely stoned honing sticks. The object of the fine sticks is to polish or lap off the tops of the original asperities to provide a larger load bearing area in order to minimize the initial 'running-in' of the surface in a tribological situation. Most plateau honed surfaces are used in a tribological process. The plateau honed surfaces are used in a tribological process. The plateau honing operation is normally associated with the production of engine bores and tends to be used only in association with the production of high quality car engines and also large industrial diesel engines. The micrograph shows evidence of adhered particles resulting from either debris or honing grits. This feature has been shown previously in relation to the grinding process.

In the case of the sample examined here, the surface exhibits clear signs of over-machining at the initial preparation. As a consequence, the operational life of the surface has been severely curtailed.

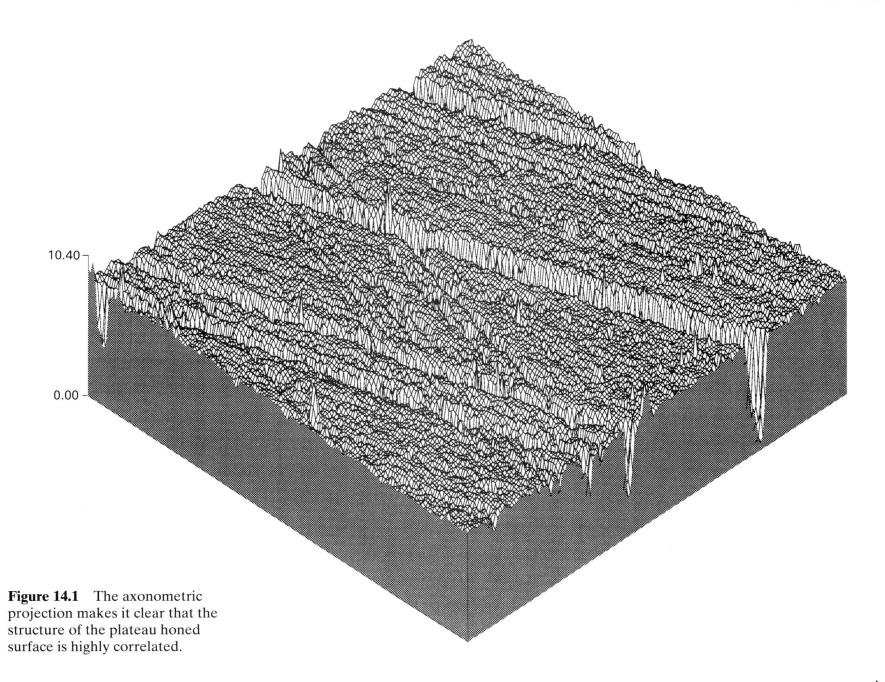

Figure 14.1 The axonometric projection makes it clear that the structure of the plateau honed surface is highly correlated.

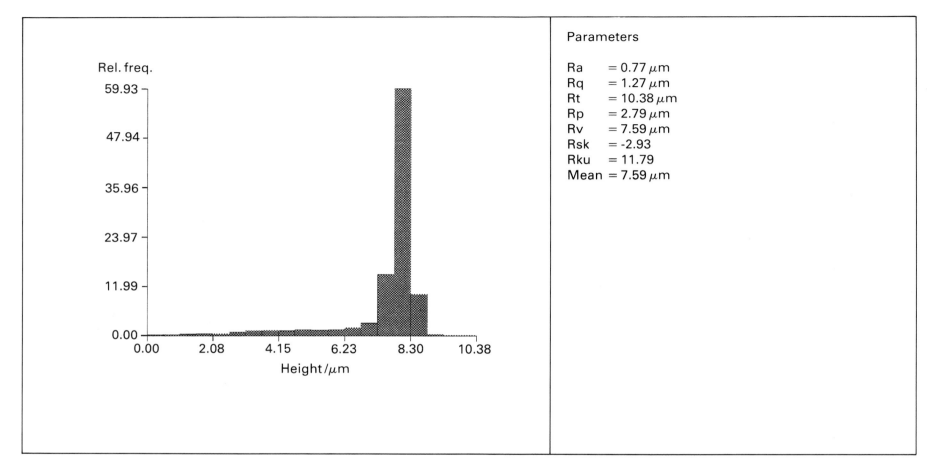

Figure 14.2 The area asperity height distribution of the plateau honed surface shows that the majority of the asperities are accumulated at, or near, the uppermost level. Notice that a plateau honed surface is negatively skewed, in this case over-skewed, at $Rsk = -2.93$. It has been established that a skewness value of $Rsk = 1.5$ would provide a surface with a bearing area suitable to provide early running-in without the production of a significant volume of debris or friction forces and without machining the surface towards early scuffing failure.

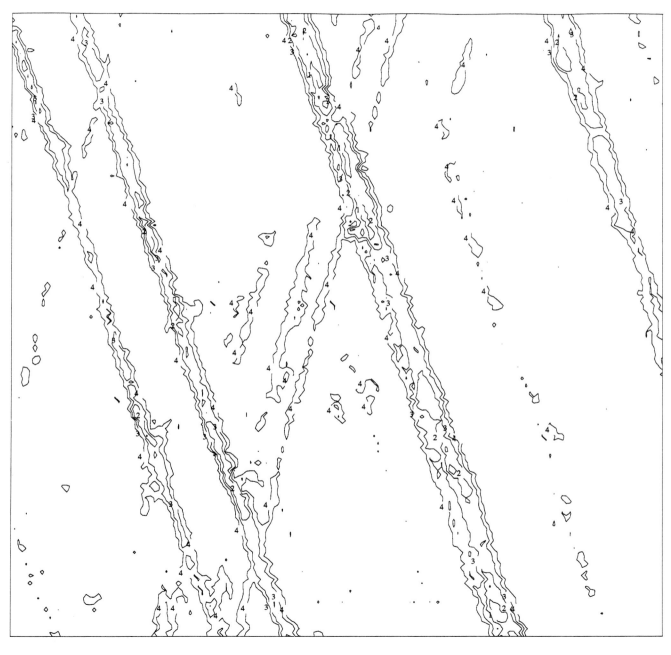

Contour key (μm)

1 : 1.04
2 : 3.11
3 : 5.19
4 : 7.26
5 : 9.34

Figure 14.3 The contour map for the plateau honed surface provides visual evidence of the oil retaining honing grooves which are associated with this machining process. These grooves are very similar to those revealed in Figure 13.3 for the honed surface.

Figure 14.4 (over) The diagram of data on 2-D profiles taken from the surface from orthogonal directions to the direction of the motion of the slider indicates that each height parameter considered yields different conditions, the measures of surface amplitude, Ra' and Rq', indicate closely similar results. The skewness and kurtosis values of the profiles show a much closer similarity than is seen for the amplitude parameters.

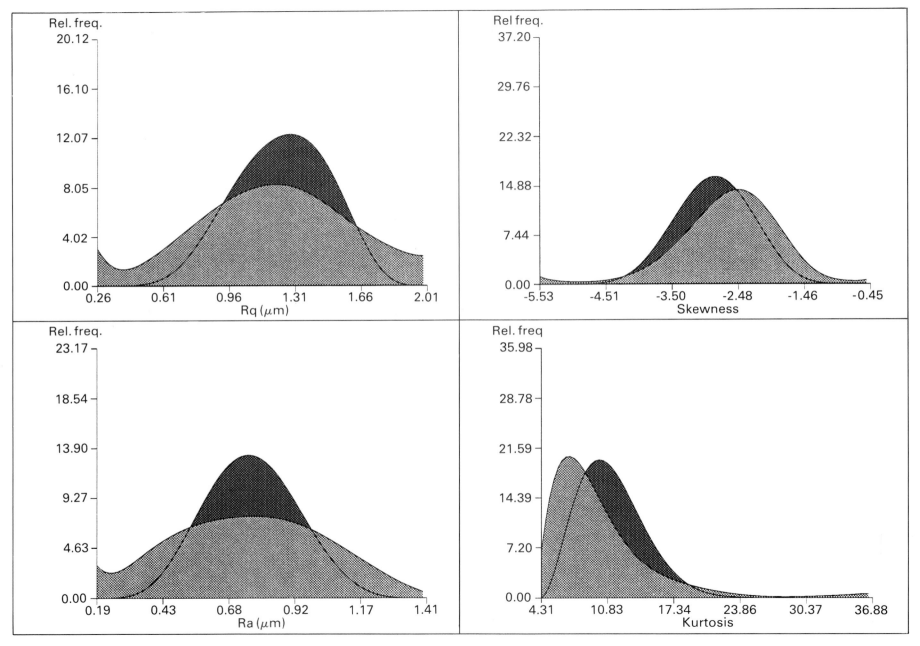

Figure 14.4

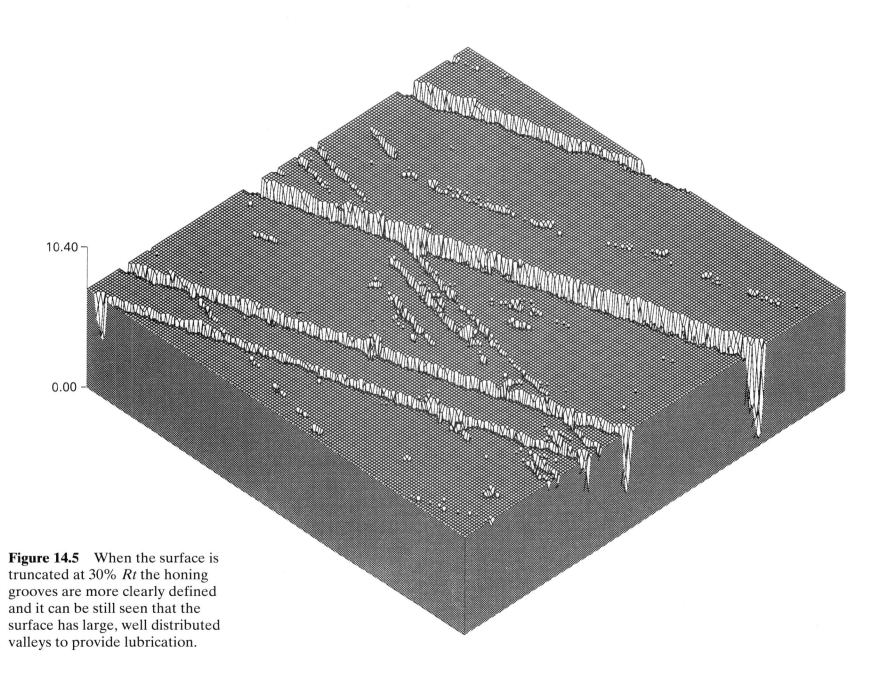

10.40 —

0.00 —

Figure 14.5 When the surface is truncated at 30% *Rt* the honing grooves are more clearly defined and it can be still seen that the surface has large, well distributed valleys to provide lubrication.

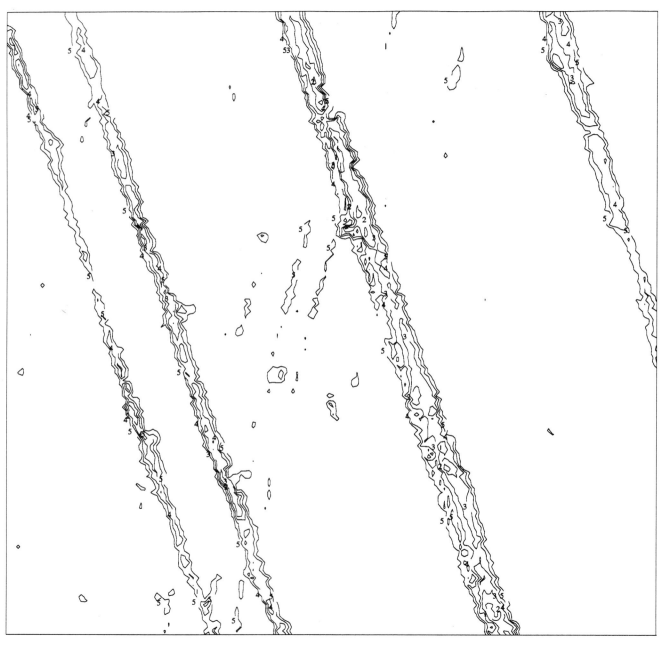

Contour key (μm)

1 : 0.73
2 : 2.18
3 : 3.63
4 : 5.08
5 : 6.53

Figure 14.6 When this contour map of the 30% *Rt* truncated surface is compared to Figure 14.3 it can be seen that some valleys have disappeared, thus reducing some of the lubricating features, but there are still sufficient valleys to ensure that the surface will operate successfully in a tribological environment.

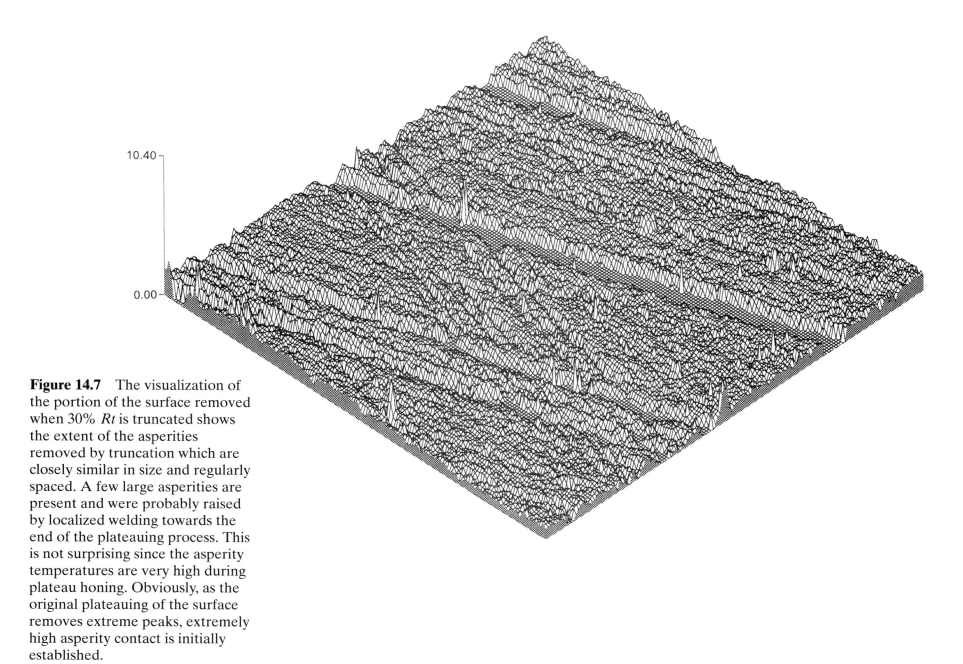

10.40 ⌐

0.00 ⌐

Figure 14.7 The visualization of the portion of the surface removed when 30% *Rt* is truncated shows the extent of the asperities removed by truncation which are closely similar in size and regularly spaced. A few large asperities are present and were probably raised by localized welding towards the end of the plateauing process. This is not surprising since the asperity temperatures are very high during plateau honing. Obviously, as the original plateauing of the surface removes extreme peaks, extremely high asperity contact is initially established.

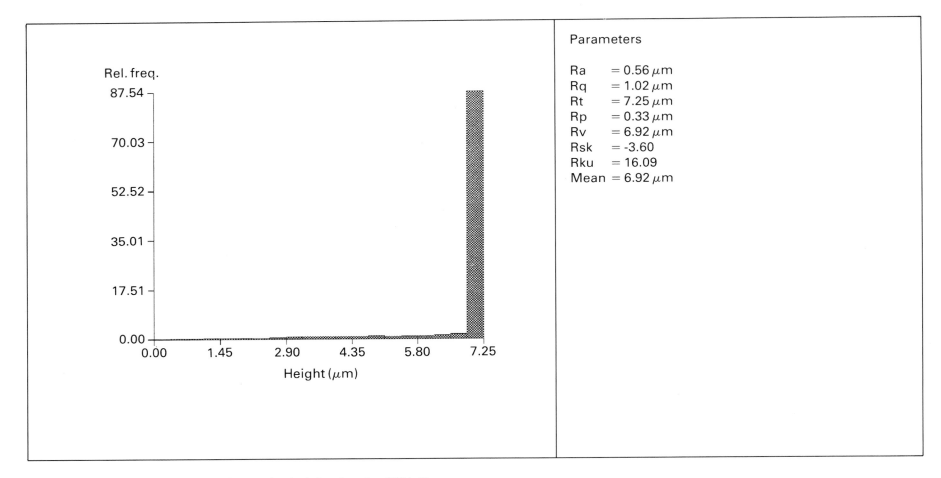

Figure 14.8 The distribution of asperity heights for the 30% *Rt* truncated surface shows that, as expected, the majority of the surface is collected at the uppermost level. As a consequence of truncation the skewness parameter has increased to *Rsk* = −3.60 and the kurtosis to *Rku* = 16.09. This implies that the surface is still tribologically acceptable although it is nearing the end of its lubrication retaining potential.

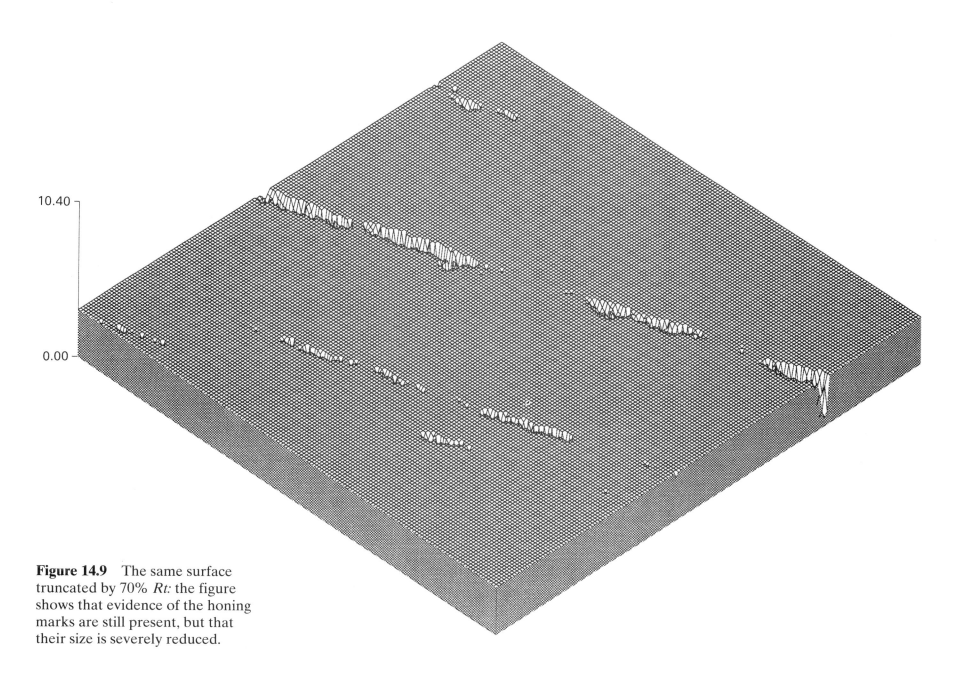

10.40

0.00

Figure 14.9 The same surface truncated by 70% *Rt:* the figure shows that evidence of the honing marks are still present, but that their size is severely reduced.

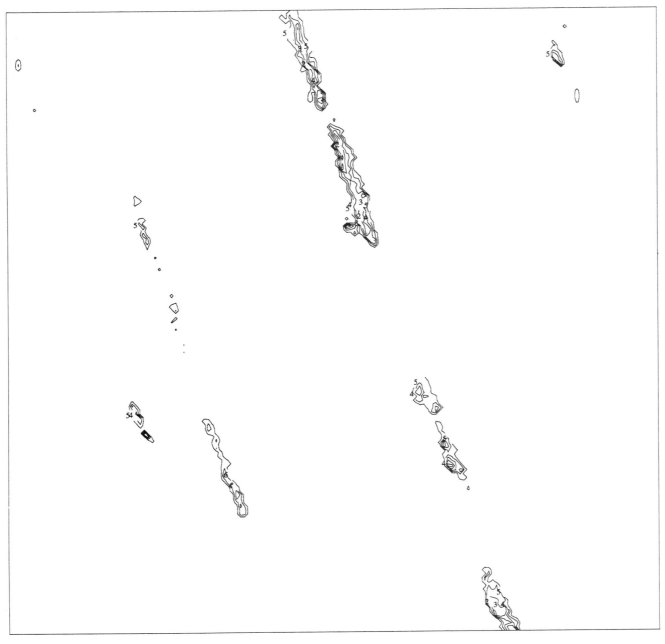

Contour key (μm)

1 : 0.31
2 : 0.94
3 : 1.56
4 : 2.19
5 : 2.81

Figure 14.10 Contour map of the surface shown in Figure 14.9. In a tribological environment, the large areas of uninterrupted plane evident here would provide an expectation of lubricant breakdown. Such a situation normally leads to scuffing and hence subsequent failure.

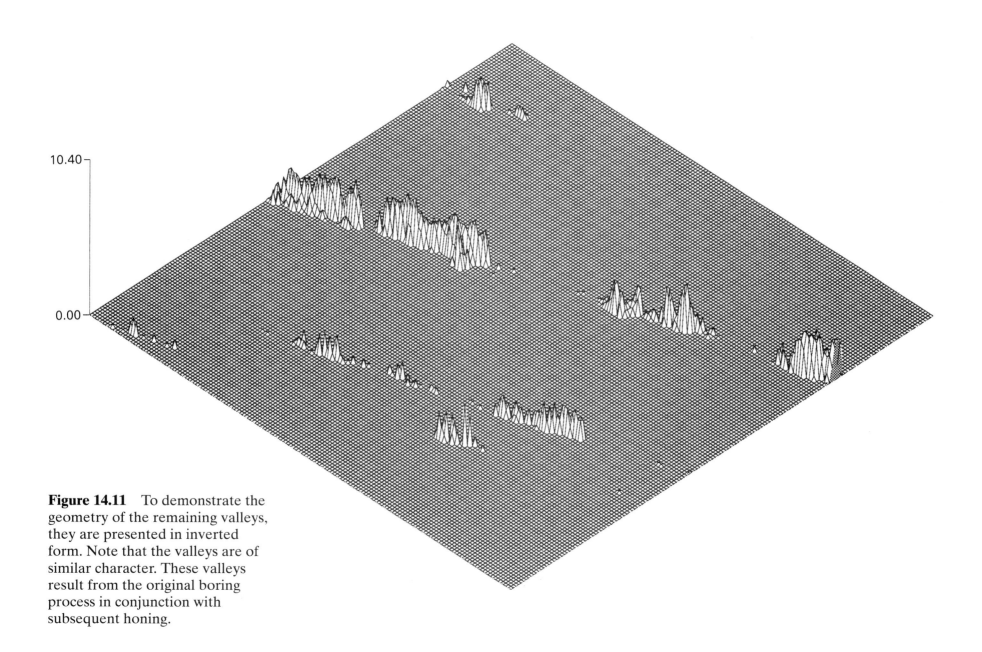

10.40

0.00

Figure 14.11 To demonstrate the geometry of the remaining valleys, they are presented in inverted form. Note that the valleys are of similar character. These valleys result from the original boring process in conjunction with subsequent honing.

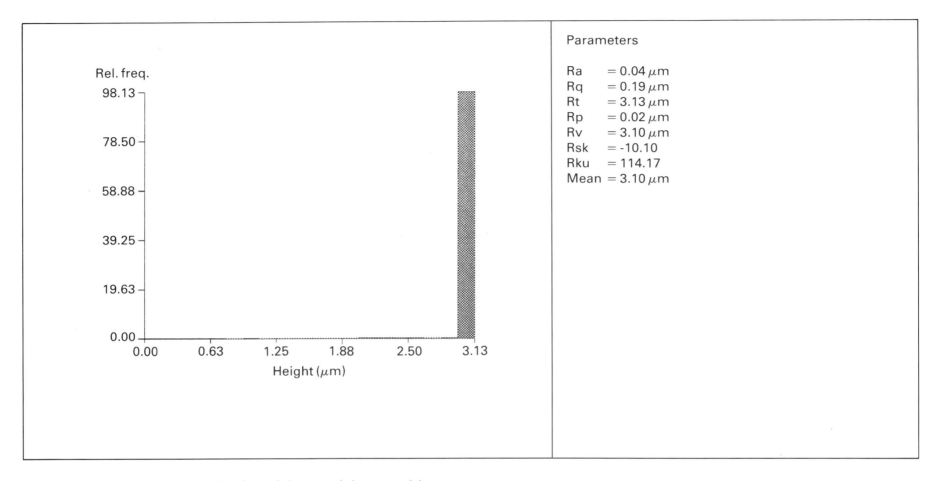

Figure 14.12   The height distribution of the remaining asperities after truncation of 70% *Rt* shows that more than 98% of the surface is collected at the uppermost level. The tribologically relevant parameter, skewness, has increased to *Rsk* = –10.10 which indicates that tribological failure is likely to have occurred. The kurtosis value of *Rku* = 114.17 supports this view.

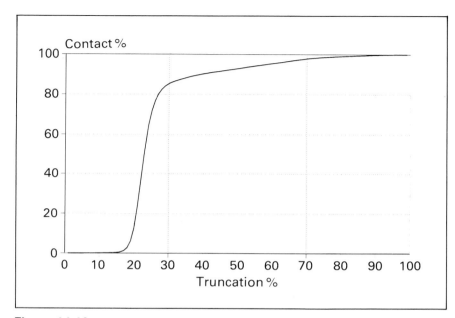

Figure 14.13

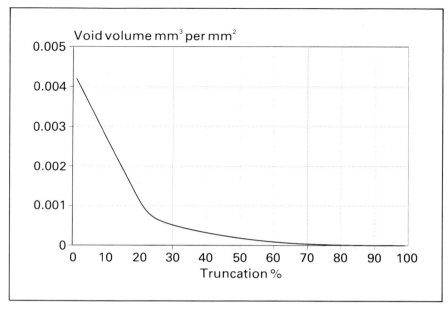

Figure 14.14

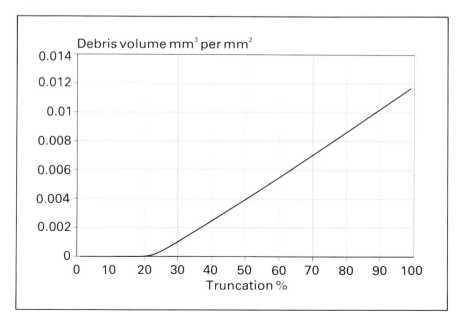

Figure 14.15

Figure 14.13   The curve of truncation % against contact % shows that the contact % approaches zero at very low values of truncation %. As the truncation reaches 20% Rt the contact % increases rapidly, reaching 86% at 30% Rt truncation. The rate of increase of contact reduces rapidly at this point. By 60% Rt truncation the contact area reaches approximately 95%.

Figure 14.14   Truncation % plotted against void volume indicates that as expected, the rate of reduction in void volume is constant until the 23% truncation level is reached. From this point onwards the volume of material being removed becomes significant and as more material is removed the void volume reduces more slowly.

Figure 14.15   The progression of debris volume information: the debris volume approaches zero at low values of truncation, up to 22%. Beyond that point, the debris volume maintains a linear increase until all the valleys are removed.

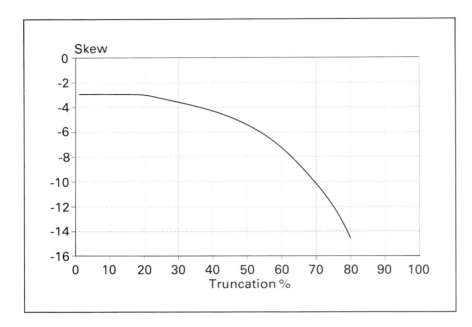

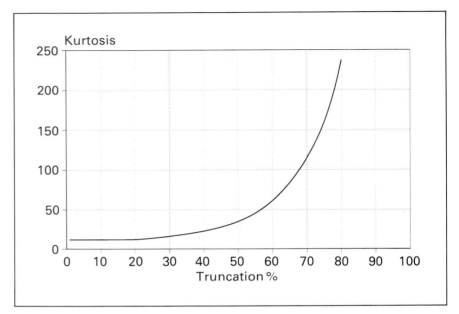

Figure 14.16   The shape parameter, skewness, has an initial value $Rsk = -2.93$ which remains substantially unchanged until the 20% Rt truncation level is reached. At further levels of truncation the value of skewness increases negatively very rapidly. When a 70% Rt truncation level is reached the value of skewness has increased to $Rsk = -10.10$.

Figure 14.17   The initial value of kurtosis remains substantially unchanged until a truncation value of 20% Rt is reached. From this point the kurtosis increases more rapidly until it becomes functionally irrelevant at the 70% Rt truncation level.

PART 2

THREE-DIMENSIONAL SURFACE ANALYSIS OF A CYLINDER BORE

INTRODUCTION

The object of this part is to demonstrate how the 3-D characterization approach given in Part 1 of this atlas may be employed to determine the potential functional behaviour of engineering surfaces.

At present, there are two methods of revealing wear within particular regions of a surface. The first method is to examine segments of the surface using a scanning electron microscope (SEM), whilst the second method is to use a mechanical stylus to map the surface in 3-D. The surface used to demonstrate the capability and scope of the stylus analysis technique is a honed cylinder bore. The object of this investigation is to estimate the long term running potential of an internal combustion engine having a surface with the characteristics as shown in Figure 1.

REAL SURFACE EVALUATION

The topography of the honed cylinder liner is shown in the axonometric projection presented in Figure 1. It can be seen that the section of the cylinder bore surface is slightly curved, as expected. In order to investigate the topography of curved surfaces it is necessary to remove this curvature mathematically. All further micrographs presented in this part of the atlas have been subject to curvature removal. Figure 2 shows the result of removing the curvature of the surface shown in Figure 1, and Figure 3 shows the height distribution of the surface.

The honing grooves are employed to hold and carry the lubricant which provides the hydrodynamic fluid film wedge to protect the surfaces during their tribological interaction. The cylinder liner used in this example has been run under typical test conditions, normally employed for lubricant assessment, on a computer-controlled test cell. To obtain data concerning the unworn topography of the cylinder bore, as it existed prior to running under the controlled test sequence, measurements were taken from the surface in a position where it could be guaranteed that running wear would not have taken place – towards the bottom of the skirt of the bore, in the region below the bottom ring travel, and in line with the position of the gudgeon pin. In this position there can be

no possibility of piston slap which can lead to mechanical damage to the surface.

The surface shown in Figure 2 demonstrates that the oil-retaining grooves which are just visible on the projection, are well spaced and lie at an included angle of 45°. This is considered by many manufacturers of cylinder liners and bores to be the optimum angle to ensure effective long-term oil lubrication and low oil consumption. The asperities appear to be well distributed over the area of the surface being assessed and are all of similar size. The average roughness value ($Ra = 0.73$ µm) is typical for a surface produced by the boring and the honing process. The negative skewness $Rsk = -2.57$ and the associated kurtosis $Rku = 21.6$ indicate that the surface has been over-machined during manufacture and this will lead to a relatively short running life. Figure 2 shows that the surface contains a number of cusps formed by the machining process; these cusps are folded over by the action of the blunt edges of the honing stones. Many of the cusps produced during the honing process will be fractured early in the three-body tribological interaction which occurs during running (the three bodies referred to are the materials of the two interacting surfaces and the lubricant). It is, however, interesting to note that some of the cusps are rigidly held on the surface and have remained in place despite the vibration and oil spray encountered during the engine running test.

To determine how the engine has performed under the action of controlled running, the liner was removed and sectioned as shown in Figure 4. Each section was assessed using 3-D stylus characterization techniques, and some more interesting sections are discussed here.

It must first be recognized that the wear life of a cylinder liner is related to:

1. the materials used in the three-body wear mechanism;
2. the lubricant;
3. the operating load and temperature conditions which prevail within the engine;
4. the operating speed of the tribological system.

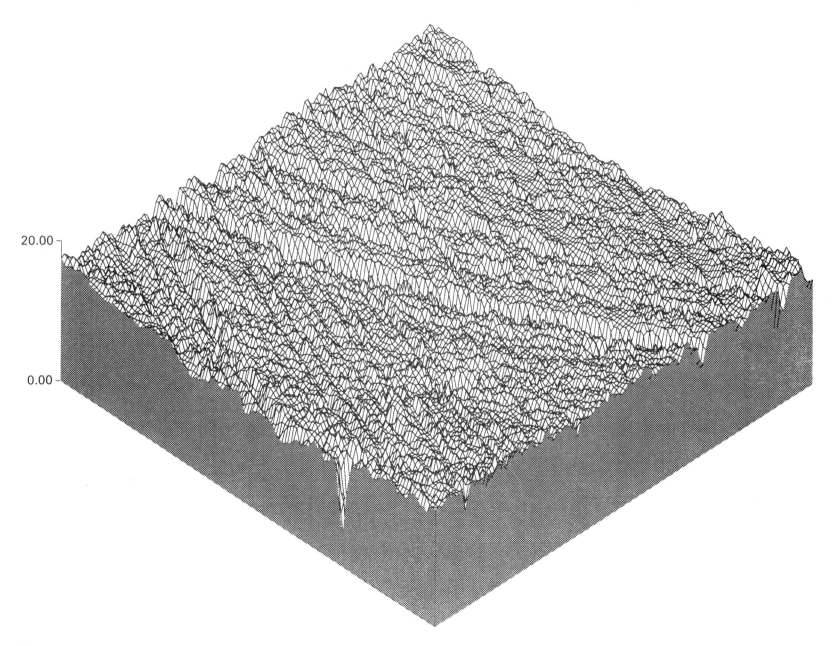

20.00

0.00

Figure 1

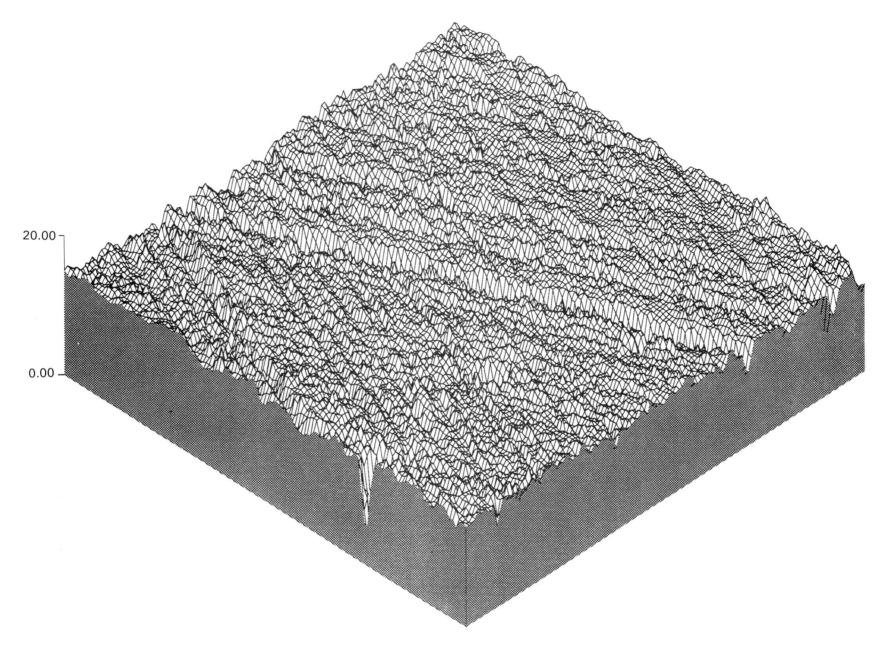

Figure 2

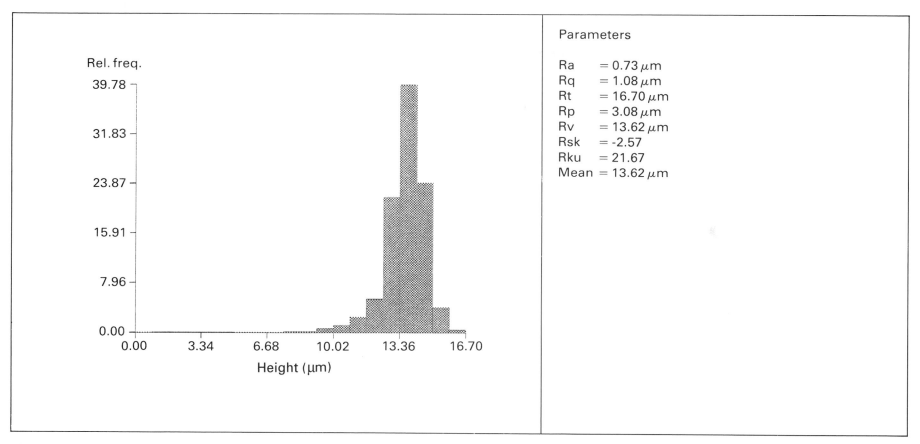

Figure 3 Height distribution and associated parameters for the surface shown in Figure 2.

The onset of failure of an engine in service is often a result of surface wear, usually bore polishing, which leads to lubrication breakdown followed by scuffing. Bore polishing can be described as a situation where, as a direct result of steady running, usually controlled wear, the oil retaining grooves and voids in the surface are reduced in size and may be subsequently removed. These oil-retaining features, as they become smaller and their separation becomes wider and the remaining grooves, begin to fail to provide enough lubricant to keep the interacting surfaces apart. The lack of lubricant starts to cause the breakdown of the hydrodynamic lubricant film and the surface scuffing mode begins.

Section 7 'run-in' region

Figure 5 presents an axonometric projection of the surface denoted section 7 in Figure 4 and Figure 6 shows the height distribution of this surface. The region shows that normal wear has taken place and that the tops of the asperities have been removed. This has had the effect of increasing the contact area at the uppermost level of the surface and reducing the surface contact

Figure 4

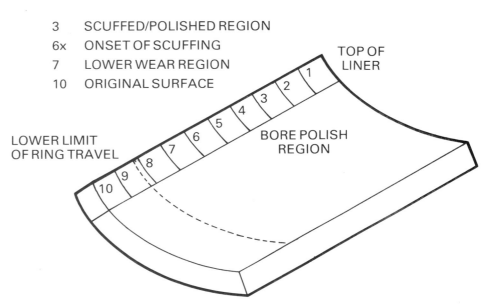

3	SCUFFED/POLISHED REGION
6x	ONSET OF SCUFFING
7	LOWER WEAR REGION
10	ORIGINAL SURFACE

pressures. The projection indicates an area of the surface which is developed as a result of the 'running-in' process. The cross-hatching associated with the honing process has become more visually pronounced and it is evident that the voids remaining are well distributed across the surface. In this region the bore roughness has decreased significantly from the original condition as expressed in Figure 2. The roughness value, Ra, has reduced by 30% and the surface has become more skewed with the value of skewness increasing by 52%. The complimentary value of kurtosis has also increased by 68%. These parameters indicate that this section of the surface can accept very little further wear before the surface lubrication starts to break down.

Section 6x 'onset of scuffing'

The region of the surface shown in Figures 7 and 8 relates to section 6x (see Figure 4) of the original bore. This region shows an interesting development, notably that the roughness value of this section of the surface has reduced quite substantially ($Ra = 0.24$).

The reduction in Ra has been accompanied by large changes in the skewness and kurtosis values. Skewness has increased to $Rsk = -6.04$ and the kurtosis value has increased to $Rku = 69.86$. Figure 7 provides evidence of very large regions which would provide poor lubrication. In addition some evidence of scuffing is now present as spikes are appearing on the surface. These spikes have been caused by asperity welding due to lack of lubrication over some regions of the surface. Figure 8 shows the height distribution of this surface.

Section 3 'bore polish'

The surface taken from section 3 (see Figure 4) is shown in Figure 9. The axonometric projection shows a further development in the surface topography, that of bore polish and the onset of surface failure. At this stage of the wear process most of the original surface has been removed; very little evidence of oil retaining features exists and most of the surface appears polished. There is significant evidence of asperities being welded and torn out of the surface; a clear example of failure by seizure. The height distribution parameters for the surface (Figure 10) have significantly changed leading to a very low value for surface roughness ($Ra = 0.10$ μm). The skewness parameter has developed to $Rsk = -6.10$ and the corresponding kurtosis value has increased to $Rku = 87.52$. The values of these parameters are consistent with surface failure. The surface shows no evidence of the original honing grooves, therefore it can be considered as having passed the point of useful life and to have failed in service. This condition is the prelude to catastrophic failure which will lead to a break up of the surface. At this point the stresses on the piston rings will become excessive, due in part to localized welding, ring damage and possibly ring 'break up' may occur.

SUMMARY OF WEAR EVALUATION OF A CYLINDER LINER

The variation in the features which have been found over the length of the bore can be accounted for by considering the variation of forces and the temperature gradients which occur within an engine during running. The force vector which acts in the plane normal to the gudgeon pin is greatest at the upper position of the stroke where the bore temperatures are highest. Also the ring velocity approaches zero at this point which reduces the possibility of producing a hydrodynamic lubricating film. As the surface topography is truncated and the lubricating reservoirs reduced, a stage is reached where the surface can no longer retain a lubricating film. The combination of high force, high temperature and lack of lubrication retention features and the presence of combustion impurities leads to metal to metal contact and subsequent surface failure when the distribution and oil retention volume of voids is too small to maintain the boundary lubricant layer. The geometry and distribution of the voids is an important consideration when specifying the surfaces for these environments.

SIMULATED SURFACE EVALUATION

In Part 1 it has been proposed that it is possible to investigate the sub-surface layers to gain an indication of the wear behaviour of a surface in a tribological situation. To demonstrate the potential of the surface truncation model, it will be applied to the honed surface originally presented in Figure 2. The following figures and paragraphs compare the surfaces modified by the truncation model to their worn form as shown previously in Figures 5–10.

Simulated 'running-in'

The original unworn surface (Figure 2) has been subjected to 'running-in' simulation using the truncation model previously discussed and the resulting modified surface is presented in Figure 11. The simulated surface shows, as would be expected, a number of flats which occur during the process of wear. When the simulated surface (Figure 11) is compared with the worn surface (Figure 5) close similarities are seen. The flats produced show very similar form, the main differences being related to the deep scratch on the original surface. It can be seen (Table 1) that the shape parameters of the surface are remarkably similar although the roughness values have a greater variation. The deep scratch is mainly responsible for this difference since a scratch was not apparent in the genuinely worn surface.

Simulated 'onset of scuffing'

The surface which exhibits evidence of the onset of scuffing (Figure 7) can be compared with its simulated counterpart (Figure 12). The two surfaces again show remarkable similarity. In both projections it is seen that evidence of the honing grooves still exists, although these are getting substantially smaller and the contact area is substantially increased. Both surfaces clearly indicate that sufficient oil retention capability still exists for both the worn and simulated areas, although the parameters for the two surfaces indicate that they are reaching the end of their useful running lives. When comparing the parameters (Table 15.1) it is interesting to note the close similarity between the shape parameters, skewness and kurtosis. The difference between the parameters may be attributed to the influence of the single deep scratch seen in the simulated surface.

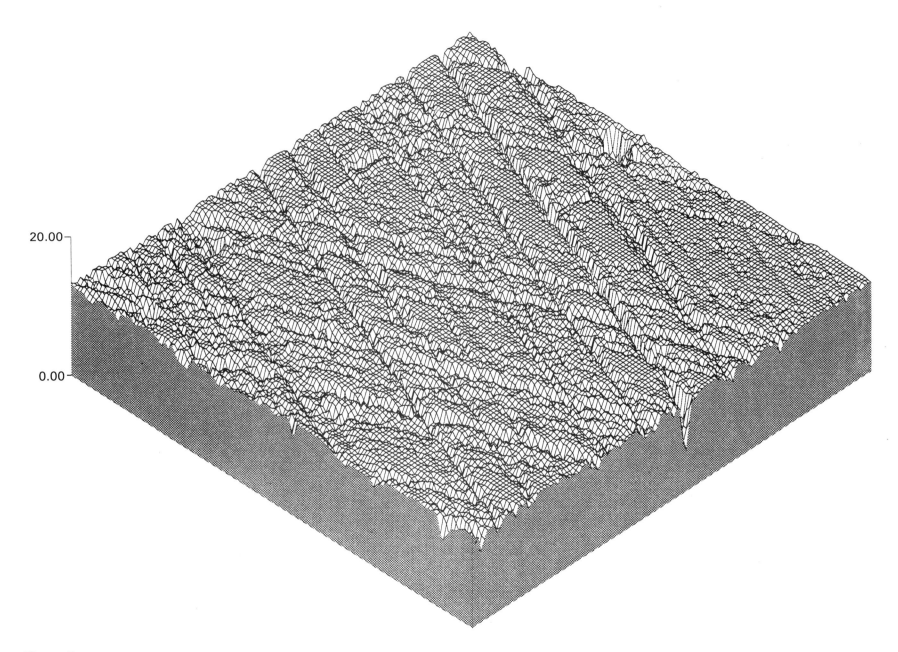

20.00 —

0.00 —

Figure 5

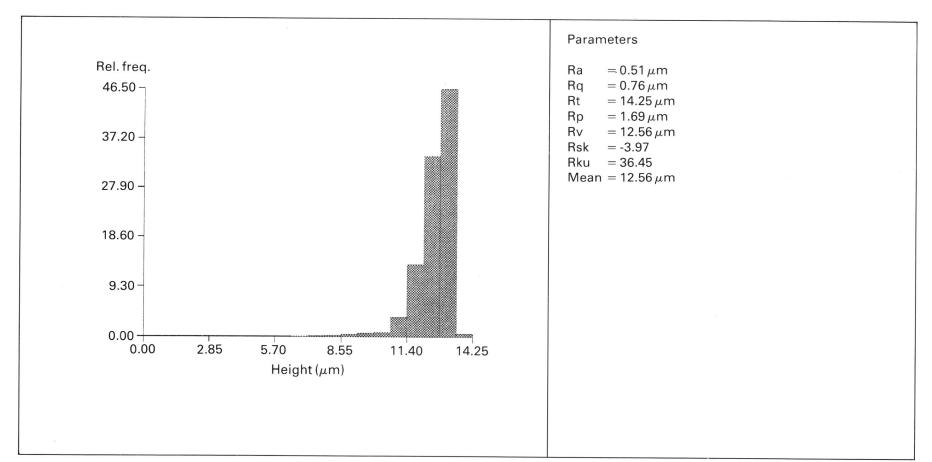

Parameters

Ra $= 0.51\,\mu\text{m}$
Rq $= 0.76\,\mu\text{m}$
Rt $= 14.25\,\mu\text{m}$
Rp $= 1.69\,\mu\text{m}$
Rv $= 12.56\,\mu\text{m}$
Rsk $= -3.97$
Rku $= 36.45$
Mean $= 12.56\,\mu\text{m}$

Figure 6

Simulated 'bore polish'

The simulation of bore polish region (Figure 13) shows more discrepancies when compared with the same region of the actual bore (Figure 9). The wear model, as it has so far been developed, does not take into account the effects of surface pick-up and, as a consequence, parameter differences are inevitable. This is clearly evident from a comparison of the figures. Also, the influence of the single valley present in the pre-simulated surface gains significance when the surface is truncated and the parameters

Table 15.1   Parameters of worn and simulated surfaces

Condition	Ra	Rq	Rt	Rsk	Rku	Figure	Comment
Worn Surface							
Run-in	0.51	0.76	14.25	-3.97	36.35	15.5	Section 7
Transition region	0.24	0.51	13.43	-6.04	69.86	15.7	Section 6x
Bore polish	0.09	0.25	7.85	-6.10	87.53	15.9	Section 3
Simulated Surface							
Run-in	0.61	0.94	14.10	-4.03	35.31	15.11	15.5% *Rt* trunc
Transition region	0.46	0.80	13.55	-5.61	57.46	15.12	19% *Rt* trunc
Bore polish	0.32	0.68	13.05	-7.55	93.03	15.13	22% *Rt* trunc

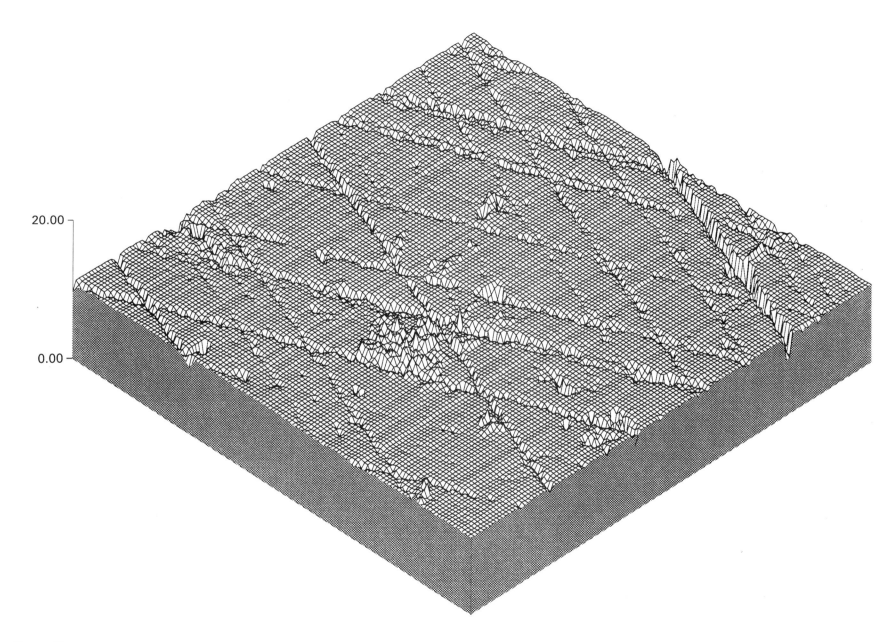

20.00

0.00

Figure 7

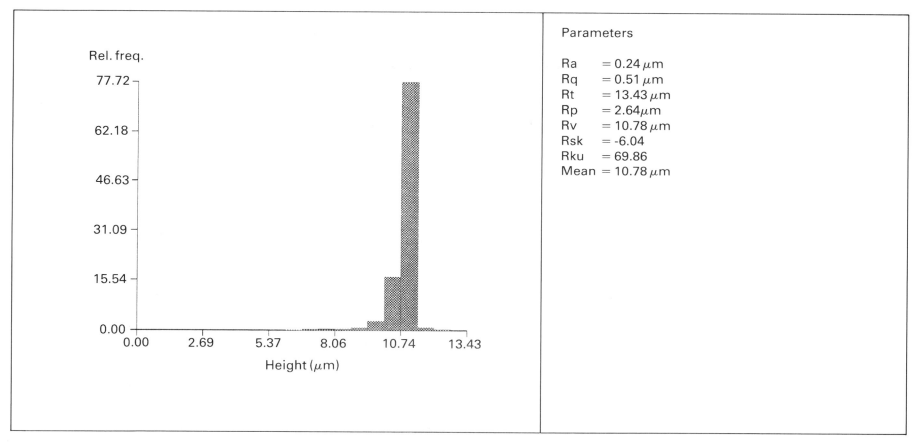

Figure 8

will exhibit increased divergence at this level of truncation. To achieve improved comparability it is necessary to develop the wear model to incorporate the topographic changes that occur due to scuffing.

CONCLUSION

In conclusion, it has been shown that truncation models can predict the likely functional performance, in a wear sense, of surfaces prepared by machining or specific finishing processs. As a consequence of such a prediction it is possible to examine the topography of the unworn surface to gain an indication as to whether the machined surface will perform during its tribological function in a way which will lead to a long operational life. In this way it is possible, early in the manufacturing sequence to reject surfaces which will not perform satisfactorily in service.

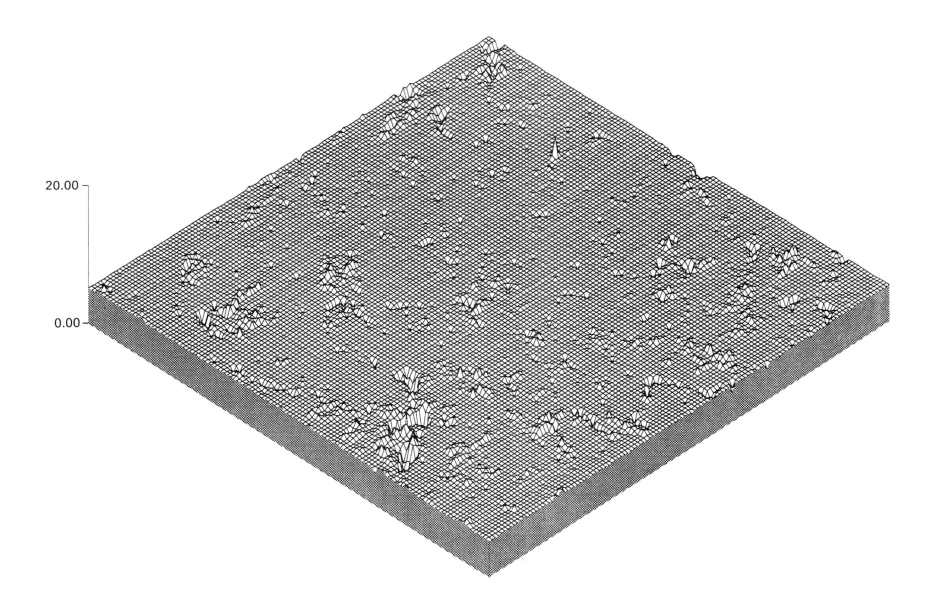

20.00 —

0.00 —

Figure 9

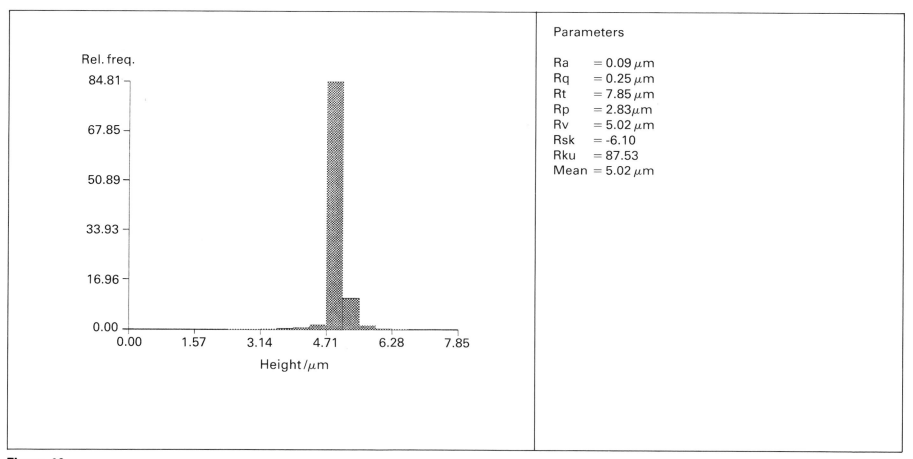

Figure 10

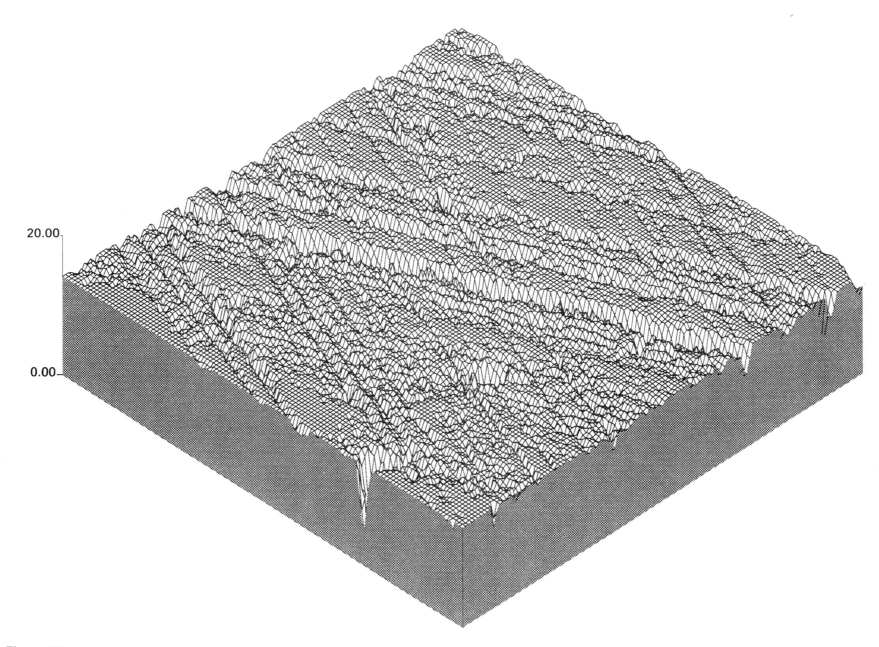

20.00

0.00

Figure 11

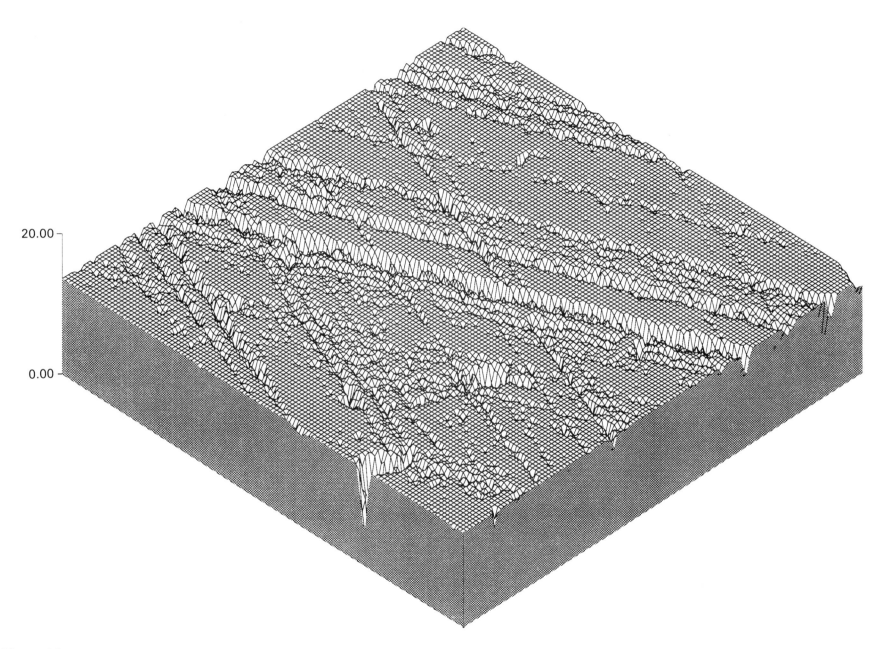

20.00

0.00

Figure 12

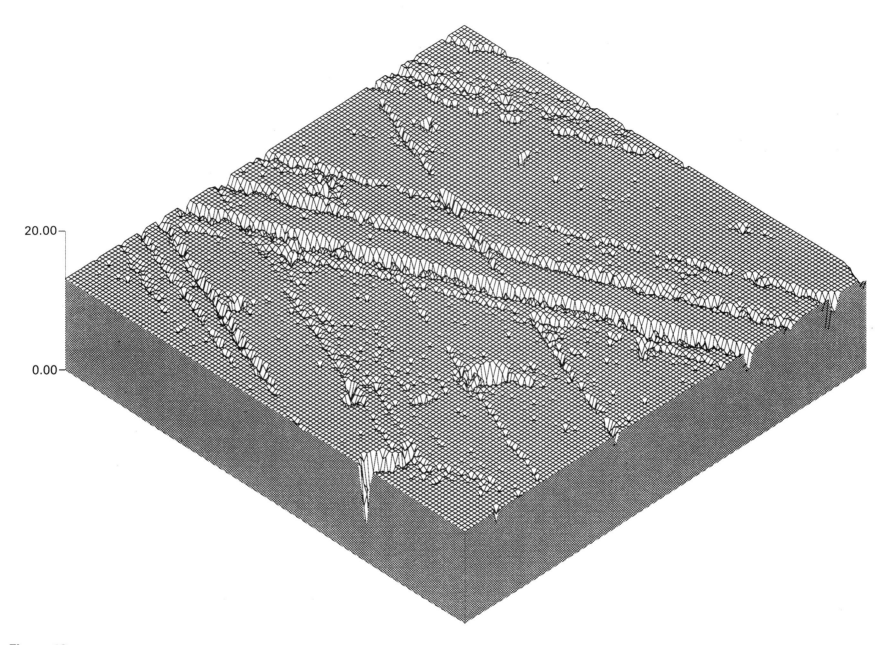

20.00 —

0.00 —

Figure 13

APPENDIX

LIST OF STANDARDS FOR SURFACE ROUGHNESS

Assessment Country	Title	Number
AUSTRALIA	The measurement of surface roughness by direct-reading stylus electronic instruments. SAA 1977	AS 1965
AUSTRALIA	Surface roughness comparison specimens. SAA 1981	AS 2382
AUSTRALIA	Surface Texture. SAA 1982	AS 2536
AUSTRIA	Measurement of surface roughness with contact (stylus) instruments. ON 1982	ONORM M 1114
AUSTRIA	Provisional surface roughness and ISO-tolerance quality. ON 1979	ONORM M 1116
CANADA	Surface texture (roughness, waviness and lay). CSA 1962	B 95
CHINA	Surface roughness, parameters and their values. CAS 1983	GB 1031 (ISO 468:1982)
DENMARK	Instruments for the measurement of surface roughness by the profile method. Contact (stylus) instruments of consecutive profile transformation. Contact profile transformation. Contact profile meters, system M (adoption of ISO 3274:1975) (E) DS 1981	DS/ISO 3274

Assessment Country	Title	Number
ENGLAND–UK	Introduction to surface texture. BSI 1982	PD 7306
ENGLAND–UK	Assessment of surface texture; method and instrumentation. BSI 1972 ISO/DIS 4287/1 (See 80/7481DC being ISO/DIS 4287/1)	BS 1134 Part 1
ENGLAND–UK	Assessment of surface texture; general information and BSI 1972	BS 1134 Part 2
ENGLAND–UK	Specification for roughness comparison specimens – turned, ground, bored, milled, shaped and planed. BSI 1974	BS 2634 Part 1
ENGLAND–UK	Specification for roughness comparison specimens – spark-eroded, shot blasted and grit blasted, and polished.	BS 2634 Part 2 (ISO 2632/11 - 1977)
ENGLAND–UK	Specification for roughness comparison specimens – Cast Surfaces. BSI 1980	BS 2634 Part 3
FRANCE	Surface texture. Measurement devices. Roughness comparison specimens. AFNOR 1981	NF E05-051
WEST GERMANY	Geometrical characteristics of surface texture. DIN 1977	DIN 4761 Draft

Assessment Country	Title	Number	Assessment Country	Title	Number
GERMANY (FR)	Surface roughness associated with types of manufacturing methods. Attainable arithmetical mean value of peak-to-valley height Rz according to DIN 4768 Part 1 (E) DIN 1981	DIN 4766 Part 1	HUNGARY	Surface roughness. Parameters and numerical values. (Supersedes MSZ 472: (S74)) MNOS 1977	MSZ KGST 638
WEST GERMANY	Determination of surface roughness values of the parameter Ra, R_2, R_{max} by means of electrical instruments stylus type. 1.DNA 1974	DIN 4768 B1.L	HUNGARY	Indicating surface roughness on technical drawings. (Supersedes MSZ15: 1975) MNOS 1979	MSZ KGST 1632
WEST GERMANY	Determination of surface roughness values Ra, R_2, R_{max} with electric stylus instruments. Basic data. DIN 1974	DIN 4768 Part 1	HUNGARY	Surface roughness. Parameters and numerical values. (Supersedes MSZ KGST 638: 1977) MNOS 1984	MSZ 4722
WEST GERMANY	Roughness comparison specimens – technical conditions of delivery – application (EV). DIN 1972 (+E)	DIN 4769 (Sheet 1)	HUNGARY	Waviness parameters. Terms and definitions, numerical values. (Supersedes MSZ 4721/3: 1975) MNOS 1981	MSZ 9655
GERMANY (FR)	Electrical contact (stylus) instruments for the measurement of surface roughness by the profile method. DIN 1979	DIN 4772	INDIA	Assessment of surface roughness. ISI 1967 + Amend 1:1974 & 2:1976	IS 3073
			INTERNATIONAL	Surface roughness. Parameters, their values and general rules for specifying requirements. ISO 1982	ISO 468
GERMANY (FR)	Measuring the surface roughness of workpieces. Visual and tactile comparison, methods by means of contact stylus instruments. (E) – DIN 1982	DIN 4775	INTERNATIONAL	Classification of instruments and devices for measurement and evaluation of the geometrical parameters of surface finish. ISO 1983	ISO 1878

Assessment Country	Title	Number	Assessment Country	Title	Number
INTERNATIONAL	Instruments for the measurement of surface roughness by the profile method. Vocabulary. ISO 1981	ISO 1879	JAPAN	Instruments for the measurement of surface roughness by the interferometric method (E). JIS 1973	JIS B 0652
INTERNATIONAL	Instruments for the measurement of surface roughness by the profile method. Contact (stylus) instruments of progressive profile transformation. Profile recording instruments. ISO 1979	ISO 1880	NETHERLANDS	Surface roughness. Measurement of Ra roughness value. NNI 1980	NEN 3635
INTERNATIONAL	Roughness comparison specimens Part 1: turned, ground, bored, milled, shaped and planed. ISO 1975	ISO 2632/1	POLAND	Surface texture: definition of terms and of CLA (centre-line average -height) index numbers (Replacing PN – 53/M-04253). PKN 1973	M. 04250
INTERNATIONAL	Instruments for the measurement of surface roughness by the profile method – contact (stylus) instruments of consecutive profile transformation – contact profile meters, system M. ISO 1975	ISO 3274	POLAND	Geometrical structure of surface. Roughness of surface. Basic definitions and parameters. PKN 1973	PN M-04251
			ROMANIA	Surface texture. Specifications of surface roughness and waviness (supersedes STAS 5730:1S66). IRS 1975	STAS 5703/2
JAPAN	Definitions and designations of surface roughness (E). JSA 1982	JIS B 0601	SPAIN	Technical drawings – method of indicating surface texture on drawings. IRANOR 1983 + Erratum	UNE 1-037-75
JAPAN	Instruments for the measurement of surface roughness by the stylus method. JSA 1976	JIS B 0651	SWEDEN	Method of indicating surface texture on drawings (E). SMS 1973	SMS 672
			SWEDEN	Guide to the choice of values for surface roughness. SIS 1975	SMS 674

Assessment Country	Title	Number
SWITZERLAND	Technical surfaces. Surface typology. Geometrical characteristics of the surface texture. VSM adopted by SNV 1976	SNV 258070 (VSM 58070)
SWITZERLAND	Surface roughness. Guidelines for using probe instruments. VSM adopted by SNV 1976	SNV 258102 (VSM 58102)
TURKEY	Surface roughness TSE 1971	TS 971
UNITED ARAB REPUBLIC	Measurement and evaluation of surface texture in finished products. EOS 1966	ES 770
USA	Surface texture. Surface roughness, waviness and lay. ANSI 1978	ANSI B46.1 (includes ANSI Y14-36 (1978))
USA	Surface Finish (RMS) SAE 1.3.60	AS 107C
USA	Surface texture, roughness, waviness. SAE 1964	AS 291D
USA	The science of ceramic machining and surface finishing. NBS 1972	NBS Sp. Publ. 348
USSR	Surface roughness. Parameters and characteristics. GKSSM 1973	GOST 2789

Assessment Country	Title	Number
USSR	State system for ensuring the uniformity measurements. Special standard and all union verification schedule for instruments measuring surface roughness parameters. $Rmax$ and Rz in the range 0.025 – 1600 mkm GKSSM 1978	GOST 8.296
USSR	Optical surface roughness measuring instruments: types, dimensions and standards of accuracy. [English Trans. (AC(MEE) 1079)] KSMIP 1961	GOST 9847
VIETNAM	Surface roughness. Parameters and characteristics. TCNN 1980	TCVN 2511
YUGOSLAVIA	Classification of surface roughness of industrial products. Explanations, terminology and definitions. JZS 1981	M.A1. 020
YUGOSLAVIA	Classification of surface roughness of industrial products. Classification of surface texture produced by machining. JZS 1964	M.A1. 023
YUGOSLAVIA	Surface roughness of industrial products. Correlation between surface quality classes and tolerance grades. JZS 1981	JUS M.A1. 025
YUGOSLAVIA	Surface roughness of industrial products. Correlation between the surface roughness and surface finishing process. JZS 1981	JUS M.A1. 026